LIVING WITH ALGORITHMS

The Information Society Series
Laura DeNardis and Michael Zimmer, Series Editors

LIVING WITH ALGORITHMS

AGENCY AND USER CULTURE IN COSTA RICA

IGNACIO SILES

The MIT Press
Cambridge, Massachusetts
London, England

The MIT Press would like to thank the anonymous peer reviewers who provided comments on drafts of this book. The generous work of academic experts is essential for establishing the authority and quality of our publications. We acknowledge with gratitude the contributions of these otherwise uncredited readers.

This book was set in Bembo Book MT Pro by Westchester Publishing Services. Printed and bound in the United States of America.

Library of Congress Cataloging-in-Publication Data

Names: Siles, Ignacio, author.
Title: Living with algorithms : agency and user culture in Costa Rica / Ignacio Siles.
Description: Cambridge, Massachusetts : The MIT Press, [2023] | Series: The
  information society series | Includes bibliographical references and index.
Identifiers: LCCN 2022021524 (print) | LCCN 2022021525 (ebook) |
  ISBN 9780262545426 (paperback) | ISBN 9780262374194 (epub) |
  ISBN 9780262374200 (pdf)
Subjects: LCSH: Artificial intelligence—Social aspects—Costa Rica. | Algorithms—
  Social aspects—Costa Rica. | Information society—Costa Rica.
Classification: LCC Q334.7 .S55 2023 (print) | LCC Q334.7 (ebook) |
  DDC 303.48/34—dc23/eng20221003
LC record available at https://lccn.loc.gov/2022021524
LC ebook record available at https://lccn.loc.gov/2022021525

10   9   8   7   6   5   4   3   2   1

Para Mariana y Marcia

CONTENTS

ACKNOWLEDGMENTS

I have benefited from the generous contributions made by many people and organizations. I could not have conducted this project without the support provided by Universidad de Costa Rica (UCR). I recognize and celebrate that a Central American public university can offer to one of its faculty members the conditions to conduct research over a relatively long period and to write a book. In one of the most vulnerable regions of the world, this is no small feat. The Research Groups Grant (*Fondo concursable para grupos de investigación*) awarded by UCR's Vicerrectoría de Investigación made it possible to collect and analyze the data. My colleagues at the Centro de Investigación en Comunicación (CICOM) provided excellent feedback and support during the research process. I thank José Luis Arce and Carolina Carazo-Barrantes for the countless hours of conversation that informed my thinking about the issues discussed in this book. I also thank CICOM's director, Yanet Martínez Toledo, and staff (María Granados Alvarado, Melissa Solano, Laura Solórzano, and Cindy Valverde) for their help at every stage of the process. Anelena Carazo offered me invaluable advice on how to improve the text's clarity and style.

I have been privileged to work on parts of this project with numerous students and colleagues over the past five years. Johan Espinoza-Rojas, Adrián Naranjo, and María Fernanda Tristán were, in many ways, the originators of this project. Their curiosity ignited my interest in this topic. Turning an idea discussed in a seminar into a research project was certainly a highlight of this process. I also had the pleasure of working with Mónica Sancho, Andrés Segura-Castillo, and Ricardo Solís-Quesada. Our collaboration is the closest experience I've had to being in a band. No algorithm could ever top their insights—or their extraordinary song recommendations! Ariana Meléndez-Moran assisted me during most of this project. I thank her for her incomparable wit and cleverness, and for her constant invitation to challenge easy explanations. I feel fortunate to work with Luciana Valerio, whose insight and creativity greatly shaped the project in its final stretch. María Fernanda Salas' curiosity and incisive questions helped refine my thinking on the issues developed in the book. Finally, I thank Melany Mora for her help with conducting several interviewees.

The arguments developed in the book benefited from comments from and conversations with numerous colleagues, including Arturo Arriagada, Jean-Samuel Beuscart, Tiziano Bonini, Carlos Brenes, Dominique Cardon, Thomas Castelain, Angèle Christin, Samuel Coavoux, Jérôme Denis, Dorismilda Flores, Katie Day Good, Erica Guevara, Mariano Navarro, Jacob Nelson, Rolando Pérez, David Pontille, Pablo Porten-Cheé, Robert Prey, Paola Ricaurte, Amy Ross-Arguedas, Larissa Tristán-Jiménez, and the Tierra Común network. I also thank Adam Fish, Rachel Plotnick, and Silvio Waisbord for their most helpful suggestions on how to improve this project and the book proposal. I feel honored to have received Pablo Boczkowski's excellent editorial advice, mentorship, and friendship over the years. His guidance was once again crucial for taking this book to completion. I also acknowledge Edgar Gómez Cruz's support, which was indispensable for finding the appropriate motivation to begin (and to finish!) the book. I thank Edgar for his faith in this project before a single page had been written.

I was also fortunate to receive insightful comments when I presented parts of this book in various venues, including the Digital Americas Conference organized by the Austrian Association for American Studies; the Centre d'Études sur les Médias, les Technologies et l'Internationalisation at Université Paris 8; GIMSSPAM at Universidad Nacional de Mar del Plata in Argentina; the Latinx Digital Media Virtual Seminar Series at Northwestern University; Universidad Nacional Autónoma de México; and Universidad Panamericana in Mexico. I presented the ideas of this book at the annual meetings of organizations such as the Association of Internet Researchers; International Communication Association; Society for Cinema and Media Studies; and Society for the Social Studies of Science. I am also thankful for opportunities to discuss the book's progress with students in seminars organized by colleagues at Heidelberg University, Northwestern University, and University of Indiana.

It has been a pleasure to work with everyone at the MIT Press. I am grateful to Gita Manaktala and Justin Kehoe for their guidance and support, and for championing this project since the very first day. I am honored to have this book included in the Information Society Series, edited by Laura DeNardis and Michael Zimmer. The production team at the MIT Press did a wonderful job in bringing the book to fruition. I also acknowledge the contributions made by other publishers who considered the book. Jonathan Gray, Aswin Punathambekar, and Adrienne Shaw, editors of the Critical Cultural Communication series at New York University Press helped me find an appropriate framework to tell the story told by this book. Anonymous reviewers of this manuscript at both MIT Press and NYU Press offered me excellent recommendations, for which I am very grateful.

Finally, I express my most heartfelt thanks to Lea and Tania, for their patience and support. Most of this book was written during periods of quarantine, social distancing, and attempts to return to "normalcy" as the world was gripped by COVID-19. I could not have completed this book without their encouragement and love. I also thank my parents, Berman and Yamileth, and sisters Marcia and Mariana, whose genuine interest in my research has been a constant source of motivation over the years.

Earlier versions of material included in various chapters first appeared as: "Folk Theories of Algorithmic Recommendations on Spotify: Enacting Data Assemblages in the Global South," in *Big Data & Society* 7 (1), 1–15 (2020), published by SAGE; "The Mutual Domestication of Users and Algorithmic Recommendations on Netflix," in *Communication, Culture & Critique*, *12* (4), 499–518 (2019), published by Oxford University Press; and "Genres as Social Affect: Cultivating Moods and Emotions through Playlists on Spotify," *Social Media + Society*, *5* (2), 1–11 (2019), published by SAGE. I thank these publishers for permission to reprint these articles.

1

DATAFICATION

What does it mean to live in a "datafied" society? Life in media-saturated contexts implies the increasing transformation of people's experiences, relations, and identities into data. Technology companies work intensively and aggressively to capture every form of human activity and turn it into data they can quantify, analyze, and profit from. Algorithms are key in this process: they seem to be everywhere and to shape people's most mundane and defining practices. An average day for many people around the world is marked by activities that require the growing use of algorithmic platforms: searching for addresses and finding a way through traffic; encountering, sharing, and commenting on the news of the day; reading thoughts and opinions from other people; finding a partner; exchanging updates with others; or consuming cultural products, such as films, series, books, and music. Algorithms have been assigned the job of determining, supposedly without any form of human bias, which are the best transit routes, the most interesting content on social media, the most compatible people in our social networks, the most relevant songs and television series, and the most germane books to read.

Today, algorithms are such an intrinsic and fundamental part of how daily life is experienced that some scholars even argue that we live in "algorithmic cultures" (Hallinan and Striphas 2016; Kushner 2013; Roberge and Seyfert 2016). This evocative notion points to the increasing difficulty of separating algorithms from the activities that make up culture. It also evinces the complex ways in which human agency and algorithmic actions are intertwined (Cohn 2019; Striphas 2015). Yet, for all the interest this notion has generated and the allure it conveys, there is an underlying paradox in its use: the study of algorithmic cultures is usually devoid of culture—in the sense given by Williams (2001, 57), that is, "particular way[s] of life, which [express] certain meanings and values not only in art and learning but also in institutions and ordinary behavior." More often than not, the analytic focus is on the algorithms themselves and how they transform subjectivity, culture, and politics (Cheney-Lippold 2017; Couldry and Mejias 2019; Fisher 2022; Schuilenburg and Peeters 2021). Algorithms are usually depicted as an external force that affects places without history and people without context. To be sure, there is much to be gained from a better understanding of how algorithms work and how they participate in conditioning people's possibilities to act and thus how algorithms transform society. But dominant approaches to the study of "algorithmic cultures" also run the risk of reproducing deterministic accounts that render people into passive victims of new forms of technological power. Although datafication is ultimately about the transformation of human life into data, most accounts of this process usually minimize or even take for granted people's own lived experiences (Livingstone 2019).

This book inverts this analytical preference by investigating how people in Costa Rica make sense of datafication processes, particularly algorithmic recommendations. To this end, I examine how Netflix, Spotify, and TikTok users in this country relate to these platforms. I study how people form personal relationships with algorithms, integrate them into the structures of their everyday life, enact them in a ritual manner, participate in public with and through them, and resist them through infrapolitical actions. In short, I look at how people in this Latin American country live with algorithms. This approach does not strive to deny the importance and consequences of algorithms but rather to emphasize the need to empirically investigate the "mutual domestication" of both users and algorithms. Before I lay out this notion and how it informed my analysis, I consider in more detail dominant approaches to the study of datafication and so-called "algorithmic power."

DATAFICATION AND ALGORITHMIC POWER

Algorithms have reached a privileged status in the study of today's digital ecology. "Meta" analyses and state-of-the-art assessments have proliferated in recent years (Cardon 2018; Dourish 2016; Kitchin 2017; Lee and Larsen 2019). As Ziewitz (2017, 2) puts it, "the question of how to study algorithms has become a topic in its own right."

Scholars have worried that algorithms have become new forms of power and capitalist domination. These concerns stem from two main types of studies. On one hand, certain researchers have looked at datafication's most significant implications and focus their studies on how technology companies extract and exploit data to shape the lives of users. Examples of these bodies of work are discussions of "surveillance capitalism" (Zuboff 2019) or "platform capitalism" (Srnicek 2016), and the growing interest in "data colonialism," which authors have defined as a form of exploitation that combines the practices of historical colonialism with new computational procedures (Couldry and Mejias 2019; 2021). In a similar manner, fears that algorithms lock users into so-called "filter bubbles" still pervade both scholarly and journalistic discourse (Bruns 2019; Pariser 2011).

On the other hand, researchers have examined how datafication works. Typically, they have focused on algorithmic platforms' operations and intrinsic logics (a process often described as "platformization") (Poell, Nieborg, and Duffy 2022; van Dijck, Poell, and de Waal 2018). These studies have also offered technological accounts of how algorithmic recommendations come into being, how they have evolved, and how big data analysis has turned into an industry (Beer 2018; Cohn 2019; Jaton 2021; Schrage 2020).

The argument for algorithmic power rests on three interrelated claims. First is the idea that algorithms carry a unique form of power that derives from sophisticated computational capacities to work with (big) data (Cardon 2018; Schuilenburg and Peeters 2021; Yeung 2018). This power is expressed through the capacity to render populations "knowable" on a historically unprecedented scale, to identify patterns in the data, and to instantiate predictive models that shape and regulate human behavior (Schwarz 2021). Through these capacities, algorithms adapt constantly to people's actions (Lee et al. 2019). As Roberge and Seyfert (2016, 3) put it, these algorithms "produc[e] numerous outputs from multiple inputs." In this sense, the power of algorithms is said to be fractal.

Second, scholars have emphasized how the power of algorithms comes from how they materialize sociotechnical networks that are difficult to disassemble (Fisher 2022; Gillespie 2016; Seaver 2019b). In this sense, algorithms can be theorized as infrastructures: they are part of larger assemblages that are gradually naturalized as an intangible and indispensable component of daily life. The most common way to frame this idea has been to compare algorithms to "black boxes" (Pasquale 2015). They are "inscrutable," argues Introna (2016). In this way, they are protected from user intervention and public regulation (Burrell 2016).

Third, authors have theorized how algorithms add a layer of complexity to broader forms of inequity in society (Benjamin 2019; Noble 2018). In addition to sophisticated computational procedures and infrastructures, algorithms intersect with larger histories of social discrimination and exploitation; they inscribe "certain ideologies or particular ways of world-making" (Bucher 2018, 30). In this way, algorithms inscribe, (re)produce, and amplify particular forms of power and hegemony in unprecedented ways.

Despite their many contributions, dominant approaches to the study of datafication and algorithms have been limited in two important ways. First, they tend to focus on either the practices of the technology companies or the platforms themselves, largely considering datafication as a process devoid of cultural dynamics and specificities. As a result, the tendency is to assume that algorithms have equal effects around the world. Put differently, research on algorithms has exemplified what Milan and Treré (2019) call "data universalism," the assumption that the use of algorithmic platforms in the global south inevitably reproduces the patterns and processes identified in other places (usually the global north or the places where algorithmic platforms are designed). Authors have suggested that, since algorithmic procedures operate in different places, their effects are identical regardless of where datafication unfolds.

Second, to date, most researchers have adopted a "top-down" perspective that privileges the study of how algorithmic structures overpower human life. Studies of datafication have focused primarily on those who extract these data to the detriment of the people who actually use algorithmic platforms. This is problematic, because, as Cardon (2018) notes, algorithmic power is not only an elusive object of study but is also assumed more often than empirically investigated.

These limitations highlight the need to examine what datafication means from the perspective of the people who use algorithmic platforms in different times and different places or, more precisely, who *enact* algorithms in their daily lives.

ENACTING ALGORITHMS IN DAILY LIFE

In this book, I argue for a more nuanced notion of datafication, one that also takes into consideration how people actually make sense of algorithms and create meaning from their lived experiences with digital platforms.

Recently, growing numbers of studies have begun to challenge the previously mentioned top-down approach to datafication. These studies don't suggest that algorithms lack "social power" (Beer 2017) but rather argue for recognizing how users enact their agency in the conditions set by algorithms.

In short, these scholars seek to understand "the agency [people] still have" in the context of datafication (Cohn 2019, 8). These studies draw on such concepts as "imaginaries," "folk theories," "stories," and user "beliefs" to explain how people become aware of and understand algorithms, and why these concepts matter for their relationships with platforms (Bucher 2018; Cotter 2019; DeVito et al. 2018; Schellewald 2022; Siles et al. 2020; Ytre-Arne and Moe 2021b). This body of work also strives to develop forms of literacies that could help people live and critically engage with algorithms (Dogruel 2021; Hargittai et al. 2020; Oeldorf-Hirsch and Neubaum 2021).

A key insight from this research is that people don't relate to algorithms in the same way. Bucher's (2018, 113) work on imaginaries—"ways of thinking about what algorithms are, what they should be, how they function, and what these imaginations, in turn, make possible"—showed that, even if algorithms are "invisible," individuals still think about and have feelings and attitudes toward them. In other words, these studies have also demonstrated that people's relationship with algorithms incorporate cognitive and affective issues (Swart 2021).

Likewise, researchers have argued that, even if users don't necessarily know how algorithmic "black boxes" operate, they relate to them and incorporate them into their daily lives through specific sets of practices, actions, and skills (Kant 2020; Klawitter and Hargittai 2018; Siles et al. 2019a; Ziewitz 2017). From this perspective, datafication and algorithmic power are not given but rather constantly made in practice (H. Kennedy 2018). Moreover, these studies have revealed how users can resist or "disobey" algorithms (Brayne and Christin 2020; Fotopoulou 2019; Velkova and Kaun 2021).

In this book, I build on these user-centered studies of algorithmic appropriation and also expand them in several ways. By putting culture at the forefront of analysis, I go beyond what previous works have done, and rather than showing that users can imagine and interact with algorithms, I provide an analysis of *why* platform users in Costa Rica relate to algorithms the way they do. Additionally, by concentrating my research on Costa Rican subjects, I contribute to the understanding of datafication in the global south. This provides much-needed balance to discussions that have focused primarily on the singularities of the United States and some European countries (Hargittai et al. 2020).

The approach I espouse in this book resonates with Latin American scholarship on the popular (*lo popular*) (Siles et al. 2022). In his book *Communication,*

Culture, and Hegemony, Martín-Barbero (1993) traced the evolution of popular classes as a historical subject since the beginning of modernity. Building on his approach, scholars have worked to turn the study of *lo popular* into a theory of communication that centers on what Martín-Barbero (1986, 284) called "the other communication," that is, how ordinary people create meaning in their situated appropriations of the media, how they "look at the world [. . .] from the experience 'of what they do with' the media" (Rincón 2015, 25, my translation). Rincón and Marroquín (2019, 44) aptly summarize this approach by defining *lo popular* as the "experiences whereby media become a part of people's daily life, and how such practices reflect submission [to] and resistance against the power, economy, and pretensions of the media's political hegemony."

Lo popular is thus a way of recognizing the agency and politics at play in people's practices of constant re-cognition (*re-conocimiento*) of themselves and their lives in the media—what Martín-Barbero (1993) called *"mediations"*. In a similar manner, this book asks not only what algorithms are doing to society but also what people are doing to and with algorithms. A focus on algorithmic "mediations" (in Martín-Barbero's sense) challenges dominant accounts of media hegemony (including datafication) that forsake the study of people's own experiences and practices, and how they recognize themselves in media technologies (including algorithmic platforms).

By considering people's practices of re-cognition in algorithmic platforms, I am not suggesting that we replace one form of agency (algorithmic) for another (human). Instead, in this book, I focus on how people *enact* algorithms in a variety of ways. The notion of enactment points to how people forge and sustain specific realities through sets of practices and relationships with heterogeneous actors (including such technologies as algorithms). Reality is a matter of practical activity rather than a single all-encompassing thing (Gad, Jensen, and Winthereik 2015; Law 2008; Mol 2002). As Law (2004, 56) puts it, "To talk about enactment [. . .] is to talk about the continuing practice of crafting. Enactment and practice never stop, and realities depend upon their continued crafting [. . .] in a combination of people, techniques, texts, architectural arrangements, and natural phenomena." Thus people don't act on assumed realities (such as datafication or algorithmic power) but rather bring them into being from moment to moment. Enactment is a complex achievement that requires constant practice and associations between people and things. To be sure, enactments don't take place in a vacuum; as Mol (2002, 10) contends, they "have a history, and they are culturally specific."

Building on these ideas, Seaver (2017) argued that algorithms are best understood as culture, that is, as the manifold consequences of practice. He summarized this idea with precision by noting that "algorithms are not singular technical objects that enter into many different cultural interactions, but are rather unstable objects, culturally enacted by the practices people use to engage with them" (Seaver 2017, 5). Multiple enactments of algorithms can thus coexist. Although the practical enactments of engineers and users can coincide, they can also differ and contradict each other without necessarily entering into confrontation (Mol 2002). Seaver (2017) contrasted this notion of algorithms as culture (or enacted through practice) with the "algorithmic cultures" approach discussed above, which posits them as an external force to culture. This distinction is useful not to deny the reality of datafication expressed by such notions as algorithmic power but rather to reframe them as practical enactments (among others). By focusing on how ordinary users enact algorithms, I seek to provide a much-needed balance to studies that have focused primarily on the practices and imaginaries of software developers and data scientists, or on the algorithms themselves.

Although my interest is in the significance of algorithms in daily life (or the domain of *lo popular*), I argue that it is impossible to disassociate how users relate to algorithms from how they appropriate algorithmic platforms. This is because these platforms are an inseparable part of "data assemblages," sociotechnical networks in which "systems of thought, forms of knowledge, finance, political economy, governmentalities and legalities, materialities and infrastructures, practices, organisations and institutions, subjectivities and communities, places, and marketplaces" mutually constitute one another (Kitchin 2014, 20). Finn (2017, 2) similarly speaks of "culture machines": assemblages of algorithms, platforms, and people. Seen in this way, platforms are the means through which people make sense of the opaqueness of algorithms.

The focus on users' enactments of algorithms is not meant to normalize or validate the corporate practices of platform companies. As discussed earlier in this chapter, scholars have done a remarkable job of making visible how algorithms perpetuate and worsen biases and existing forms of inequality (Eubanks 2018; Wachter-Boettcher 2017). But, once again, work on these issues has tended to focus on the operations of algorithms themselves.

Accordingly, a preferred method has been to "audit algorithms" to detect and prevent discrimination (Sandvig et al. 2014). Despite its importance, most work on these issues tends to take for granted users' motivations and practices without actually studying them. In some studies, an underlying assumption is that users have "little ability to impact the algorithm" (Noble 2018, 179). I argue for turning this assumption into an empirical question. A focus on how users enact algorithms could help to nuance and broaden our understanding of the operation of algorithmic power in daily life. From this perspective, the study of sexist and biased algorithms would significantly gain from a critical consideration of how people in specific places and times make sense of, experience, and enact datafication processes.

MUTUAL DOMESTICATION

The study of algorithmic power and the agency of users has typically been framed in oppositional terms: either algorithms have "power" or users have "agency." As an alternative, I situate contemporary debates about the role of algorithms within a logic of "mutual domestication": algorithms are designed to turn people into ideal consumers for data extraction purposes, but users enact these algorithmic recommendations in particular ways as they incorporate them into their daily lives. In this book, I explore five dynamics through which mutual domestication occurs: *personalization* (the ways in which communication relationships between users and algorithmic platforms are built); *integration* (how algorithmic recommendations are combined in a matrix of cultural resources); *rituals* (how users enact the centrality of algorithmic platforms through patterned actions in their daily lives); *conversion* (the transformation of private relationships with algorithms into a public issue); and *resistance* (how people challenge various aspects of algorithmic platforms). I develop this argument by privileging the study of cultural dynamics rather than that of the platforms and their alleged intrinsic capacities. In other words, this book looks across the use of three algorithmic platforms (Netflix, Spotify, and TikTok) to provide a nuanced and rich account of the cultural dynamics that shape datafication.

Borrowing a phrase from Ian Hacking's (1999, 35) classic analysis of the "social construction" metaphor, it is necessary to clarify "what work [domestication] is doing." In media studies and communication studies, the term owes much of its popularity and depth to the work of Roger Silverstone and collaborators. For Silverstone (1994, 98), domestication meant

> the capacity of a social group (a household, a family, but also an organisation) to appropriate technological artefacts and delivery systems into its own culture—its own spaces and times, its own aesthetic and its own functioning—to control them and to render them more or less 'invisible' within the daily routines of daily life.

Despite the emphasis on the work of audiences in domesticating television, Silverstone did not neglect the importance (or power) of media technologies. Silverstone (1994, 4) himself worried about the possibility of television's "*coloni[zation]* [of] basic levels of social reality" and argued for the need to study people's domestication practices to further understand forms of technological power and thus avoid "mistak[ing] the basis of [this] power and misjudg[ing] the difficulties in changing or controlling it" (emphasis added). Thus, although similar to the notion of "mutual shaping" (Boczkowski 1999), I prefer to speak of mutual domestication to highlight issues of meaning making rather than focusing on how users try to transform algorithms or the system of datafication that subtends their operation. Silverstone (1994, 98) articulated similar ideas in his use of the term domestication:

> Domestication [. . .] is an elastic process. It stretches all the way from complete transformation and incorporation to a kind of begrudging acceptance, and from total integration to marginalisation. But what links both extremes is the quality of the work involved, the effort and the activity which people bring to their consumption of objects and their incorporation into the structure of their everyday lives.

More than a straightforward application of Silverstone's domestication theory to the case of algorithms, this book follows his broader interest in analyzing the dynamic ways in which people relate to artifacts and media content in their daily lives. I also take inspiration in Silverstone's underlying effort to turn the study of people's consumption practices into a theoretical intervention. By focusing on domestication issues, Silverstone sought to counterbalance two dominant theoretical extremes: on one hand, the insistence of the Frankfurt School on the overdetermination of cultural industries and, on the other, an idealization of media consumers. The notion of mutual domestication helps balance out a similar dichotomy in which current discussions about datafication and the role of algorithms now take place (that is, the notions of algorithmic power and human agency).

In *Television and Everyday Life*, Silverstone (1994) argued for a dialectical approach. He explicitly proposed a "synthesis" between Marshall McLuhan's and James Carey's views of communication (Silverstone 1994, 94). He also recognized the centrality of Giddens' structuration theory in his understanding of domestication. This link to Giddens helps frame the relationship between users and technologies as a matter of "duality" (Webster 2011): technologies are both enacted by people through practices and institutionalized in specific structures (Orlikowski 1992); they are both the product of user practice and condition (without determining) human action. (I return to the links between mutual domestication, Giddens' theory of structuration, and the notion of enactment in chapter 7.)

DATAFICATION AS CULTURE

The question with which I began this book has a long history in the social sciences and humanities, most notably in media and communication studies. In the mid-1990s, Ien Ang (1996, 61) asked: "What [does it] mean or what it is like to live in a media-saturated world[?]" This question stemmed from a concern about a lack of sufficient attention to television audiences. A few years later, Nick Couldry framed the very same question in relation to "a society dominated by large-scale media institutions" (Couldry 2000, 6). Couldry thus stressed the need to examine the practices of what he considered to be a neglected research group: "ordinary people." In a similar manner, there has been a historic tendency to emphasize the need to study "what people do with the media" rather than the other way around. This exact phrase was used by researchers who wrote about uses and gratifications in the 1960s, audience activity in the 1990s, and practice theory in the 2000s, to name just a few topics of study.

These patterns point to a chronic tendency to privilege deterministic accounts of the significance of media technologies in media and communication studies. In many ways, the debates about datafication rehash old concerns about the power of media and technology. C. W. Anderson (2021) recently explored some of the factors that might account for an apparent return to the notion of powerful effects in communication and media research regarding algorithmic platforms. Anderson posited three main reasons to explain this state of affairs: an external system of rewards and funding that legitimizes certain themes and approaches; an imprecise interpretation of communication studies' history and main conclusions about the influence of the media in society; and the performativity of the methods used by data science (which tend to produce certain kinds of questions and conclusions). In addition to Anderson's points, it is reasonable to suggest that historically, communication studies have been more inclined to study artifacts and objects rather than people (Waisbord 2019). There also seems to be a fascination with the sophistication of technological companies' operations fused with the concerns these operations have raised. Livingstone (2019, 176) arrives at a similar conclusion when she argues that "in accounts of the datafication of society, attention to empirical audiences is easily displaced by a fascination with the data traces they leave, deliberately or inadvertently, in the digital record." This displacement could be precluding the epistemic break that is required to problematize the influence of these companies in public discourse, culture, and scholarship.

Some observers have claimed that datafication is something of a beast of a different nature (Danaher 2019). For example, Milan (2019, 220–221) contends that "datafication by nature is far more pervasive and surreptitious in its mechanisms than what we have known historically as 'media'." I argue for turning such premises about the historical singularity of datafication into empirical products. Despite the apparently "surreptitious" nature of datafication, history could also teach us a few lessons about how people have related to technologies that seemed to overpower their lives and how intellectually productive it has been to explore the uses of the media as a lived experience. Some of the most fruitful approaches in the study of the media came precisely as reactions to claims of technological power. Carey's (1992) seminal work, *Communication as Culture*, provided an alternative explanation of the media–culture relationship by inverting common assumptions on the meaning of communication. I take inspiration from this work to argue for the need to understand datafication as culture, a lived experience through which realities are produced, maintained, and transformed in Latin America. That is the task set out for this book.

ALGORITHMIC PLATFORMS IN LATIN AMERICA

Latin America offers an ideal site for a study of this kind. Algorithmic platforms have experienced an extraordinary growth over the past years in the region. According to recent reports, Latin America accounts for more than 16 percent of Netflix's profits and 20 percent of Spotify's annual revenue (Statista 2021a; 2021b). I chose Netflix, Spotify, and TikTok for this study because of the centrality of algorithms in their technological and economic models (Bandy and Diakopoulos 2020; Eriksson et al. 2019; Lobato 2019). Furthermore, these platforms have different trajectories in Latin America, which makes it possible to examine how people relate both to established technologies and those whose social meaning or identity is not yet entirely stable. This strategy also allows for the study of algorithmic platforms that, while similar in their focus on entertainment, differ in the kind of content that is recommended: series and movies, music and podcasts, and user-created videos and sounds.

NETFLIX

Early in 2011, Netflix announced plans to launch its streaming services in forty-three Latin American and Caribbean countries. After Canada, Latin America was the first region where Netflix became available outside the United States. According to Lobato (2019), this decision was justified because of the region's relatively large middle class, infrastructural conditions, and familiarity with the concept of pay TV. Although it was the first US company to offer streaming services, other options, such as Telefónica's On Video, already operated in some Latin American countries. In September of that year, Netflix launched the platform with the same price it charged in the United States and Canada (approximately US$7.99 per month for its basic plan). This pricing ensured that the service was more accessible for the elites and upper-middle classes of the region (Lobato 2019; Straubhaar et al. 2019). As a point of comparison, Costa Rica's monthly minimum wage in 2021 was almost $500, but studies have shown that a significant number of workers in both private and public sectors earn less than $400 per month (Pomareda 2021). Moreover, the cost of living in the country is relatively high. Costa Rica is among the most expensive countries in Latin America (Mata Hidalgo et al. 2020).

Three main concerns surrounded Netflix's launch in Latin America. First, the use of software and websites that allowed free access to content (such as Cuevana and later on Popcorn Time) was common in the region. By the time Netflix launched, Cuevana was already among the most accessed websites in some of the region's countries (Varise 2011). Second, a relative lack of sufficient broadband could slow down the evolution of streaming services (Jordán, Galperin, and Peres Núñez 2010). Despite significant growth, studies at the time noted that "the region as a whole continue[d] to trail behind international best practices in leveraging ICT advances" (Dutta and Mia 2011, 25). Third, some commentators noted that, compared to other parts of the world, the region lacked a robust culture of online credit card payment (Rubio 2012).

According to company representatives, Netflix's start in Latin America was "challenging" (Rubio 2012). The company kept its mission to guarantee that all of its content would include Spanish subtitles; it also continued to work with local banks to widen the range of options available to pay for its service (such as PayPal). In 2013 and 2015, the company launched a series of contests in Brazil and Mexico, respectively, and asked users to vote for a series of films produced in these countries. The winners were included in the platform's catalog for some time. These three strategies proved successful. By 2015, Brazil and Mexico were already among the countries with the most Netflix subscribers, after the United States, Canada, and the United Kingdom (Ribke 2021; Siri 2016).

Netflix's strategy to use local content to strengthen its position in Latin America as part of a larger global plan to increase the number of original content works (as opposed to licensed works) was quickly revealed. The next step in the strategy of solidifying Netflix's position through local content was to produce original series in countries of the region. Between 2015 and early 2017, Netflix released its first four shows produced in Latin America: *Club de Cuervos* (Mexico), a comedy about a family that owns a local league soccer club; *Narcos* (Colombia), the story of Pablo Escobar and the rise of drug cartels in this country; *3%* (Brazil), a dystopian thriller set in a near-future Brazil in which some people compete to join a selective society; and *Ingobernable* (Mexico), which stars Kate del Castillo (a well-known actress who started her career in Mexican *telenovelas*) as the country's first lady. With this group of early productions, Netflix thus sought to capitalize on the history of established television genres (such as *telenovelas*) and themes in the region (like soccer, politics, and drug trafficking) while at the same time developing series that could also be appealing to audiences abroad (Piñón 2014). For this reason, Ribke (2021, 176) refers to Netflix's content-development strategy in the region as a "pan-Latin American fiction formula."

While Netflix's algorithmic model of content development and recommendation has fascinated researchers in the global north, it has received comparatively much less attention in Latin America (Finn 2017; Hallinan and Striphas 2016). Latin American scholars have privileged instead the study of the intersection between particular genres with a long history in the region and the platform's overall development (Gómez Ponce 2019; Orozco 2020; Ribke 2021). For some researchers, Netflix has contributed to a new aesthetics of violence through its treatment of drug cartels in shows like *Narcos* (Amaya Trujillo and Charlois Allende 2018; Blanco Pérez 2020; Cozzi 2019). Others have focused on the platform's strategic use of melodramas, the quintessential genre in Latin America's audiovisual popular culture (Cornelio-Marí 2020; Higueras Ruiz 2019; Llamas-Rodriguez 2020; P. J. Smith 2019).

Netflix's plans to develop original content in Latin America continued "quickly" and "aggressively" over the next few years (Astudillo 2017). Reed Hastings, Netflix's CEO, noted that, if the platform's beginning in Latin America had been relatively "weak," by 2018 it was "a rocket" (cited in M. Dias and Navarro 2018, 19). During an event named *Vive Netflix* held in Mexico City in 2017, Ted Sarandos, CEO of Netflix, did not hesitate to call Latin America the "most dynamic" and "most important" region for Netflix outside the United States (AFP 2017). Netflix executives spoke publicly about the "need" to produce content for countries beyond Brazil and Mexico. Said Erik Barmak, Netflix's vice president of international original content at the time: "We discovered that we had to be active in all the big markets, not only Brazil and Mexico but also Colombia and Argentina" (Ordóñez Ghio 2017). This shift took place during a period of important development of broadband infrastructure in Latin America. As a report from the United Nations Economic Commission for Latin America and the Caribbean (ECLAC 2017) shows, the number of households connected to the Internet in the region grew by more than 100 percent between 2010 and 2016, and mobile broadband reached 64 percent of the population in 2016.

By 2018, Netflix announced it had fifty projects in different stages of development in Latin America, including series, films, documentaries, and stand-up comedies (Netflix 2017). This strategy has continued over the years, either through the development of local content or exclusive licensing deals for the region. Importantly, the production of content in Latin America has had a clear pattern: it has concentrated in countries with the largest user markets and more consolidated audiovisual industries, namely, Argentina, Brazil, Colombia, and Mexico. Moreover, as Penner and Straubhaar (2020) contend, Netflix's approach to local production in Latin America has not altered the region's traditional dependence on cultural products from the United States in any major way. In practice, Netflix has perpetuated the unilateral flow of content from the United States and a few other countries to the rest of the world (Aguiar and Waldfogel 2018).

By the end of 2019, Netflix had almost 30 million subscribers in Latin America, 22 percent more than the previous year (Solsman 2019). It was the most used streaming platform in the region and represented approximately 16 percent of the company's revenues (Ribke 2021). According to Ribke (2021, 165), "for a region that barely represents 6% of the world's GDP, Netflix's income from Latin America is stunning."

Latin America has not been exempt from the intense fiscal challenges the platform faces in many parts of the world. For example, Dias and Navarro (2018) traced the early opposition that Netflix has faced from Brazilian organizations that have accused it of unfair competition and demanded that it pay more taxes. The issue of taxes became a heated subject of legislative debate across Latin America (Herreros 2019). According to an ECLAC report, by 2019, countries like Argentina, Brazil, Chile, Colombia, Costa Rica, Mexico, Paraguay, and Uruguay had either begun to collect value-added tax (VAT) for digital platforms such as Netflix (and Spotify) (approximately 20 percent in most countries) or were about to do so at the turn of the decade (CEPAL 2019).

In this context of steady user growth, the rise of competition from other streaming services, and increasing fiscal pressure, the company has indicated that it will continue to invest in the production of original content from the region's largest user markets, most notably Brazil and Mexico (Hecht 2019).

SPOTIFY

Spotify launched in Latin America between 2013 and 2014, precisely at the time its identity was shifting from music provider to an algorithm-powered music recommender (Eriksson et al. 2019). Early in 2013, Spotify acquired Tunigo, a music discovery app that allowed users to find playlists based on particular activities and moods. In 2014, it made another key acquisition: an algorithmic recommender platform named The Echo Nest. Eriksson and colleagues (2019) refer to this transition in Spotify's identity as the "curatorial turn." With an exponential increase in venture capital investment, from 2013 to 2015, the company expanded to new territories, including seventeen countries in Latin America. Spotify representatives emphasized similar ideas put forth by Netflix to account for this decision: in addition to vibrant music-production cultures, the region had a relatively large middle class, and an important segment of the population was young, which seemed ideal for embracing mobile apps (Nicolau 2019). As in other parts of the world, in Latin America, Spotify's free version came with ads. The monthly cost of the ad-free ("Premium") subscription slightly varied across countries but was cheaper by a few dollars than in the United States ($9.99 a month). (As noted above, this represents 2 percent of the legal monthly minimum wage in Costa Rica.)

By the time Spotify launched in Latin America, the region had "one of the highest rates of digital piracy in the world" (Yúdice 2012, 18). Various reports estimated that approximately 33 percent of Latin America's population had access to mobile Internet at the time (CEPAL 2014; GSMA 2016). Although streaming services (such as Deezer) were already available in the region, the majority of Latin Americans obtained their music through informal markets (burned CDs and MP3 downloads) (Ávila-Torres 2016; Karaganis 2011; Pertierra 2012). Spotify representatives strategically commented on this issue in the media, framing it as the "problem" to which its platform provided a solution. Thus, Gustavo Diament, Spotify's managing director for Latin America at the time, argued early in 2015 that the company's biggest rival in Latin America was not another streaming platform but rather "piracy" (Pineda and Morales 2015).

To deal with the difficulties of convincing users to pay for a monthly subscription with a credit card (a problem also faced by Netflix), Spotify opted to deal directly with telecom operators in Latin America. With this strategy, Spotify sought to include the app as part of users' mobile plans and thus counteract the relative advantage afforded by Apple's position in the mobile market (Fagenson 2015). Finally, Spotify also opened an office in Miami with the primary purpose of strengthening its position in Latin America. According to company representatives, Miami offered the company multiple "pan-regional opportunities," mobile advertising possibilities, and a chance to influence the Latinx market in the United States (L. Martínez 2015).

Spotify's strategies produced relatively fast results. Unlike Netflix's uptake, which its executives defined as "challenging," Spotify representatives asserted that the success of the platform in Latin America had involved "almost no efforts" on their part (Nicolau 2019). Within a year of its launch, Mexico was already among Spotify's five biggest country markets. The company expected that this country would be among its top three user markets by the end of 2015 (Pineda and Morales 2015). In 2018, 20 percent of the platform's paying subscribers were in Latin America, a percentage that has remained stable over the years (Fortune 2018). Spotify's financial results for mid-2018, the first report after the company's IPO, stated that the "growth in [. . .] emerging regions of Latin America and Rest of World continue[d] to out-pace growth in more established markets" (Business Wire 2018). In addition to Mexico, Miami, and Brazil, throughout 2018 and 2019, Spotify opened small offices in countries with relatively large consumer markets (namely, Argentina and Colombia).

Spotify representatives have emphasized various themes in the media to comment on the platform's solid growth in Latin America. They have often suggested that Latin America has become a "special" region for the company. Mia Nygren, Spotify's managing director for Latin America in 2019, thus analyzed the platform's success in Mexico: "Mexico has been special for us. It was the first country where we launched in Latin America. And if you add the intensity with which the Mexican user consumes music and the number of hours they spend on the platform, we have a perfect love story" (Hernández Armenta 2019, my translation).

Nygren's discourse puts forth the idea of a "perfect match" between some supposedly essential traits of music consumers in countries like Mexico and the technological affordances provided by Spotify (most notably its algorithmic personalization and multiplatform ubiquity) (R. Martínez 2019). Because of the number of users and artists in the city, Spotify began employing the notion of "Streaming Capital of the World" to refer to Mexico City. Company representatives also started to showcase Mexico in particular, and Latin America in general, as the model of success they wanted to emulate in other parts of the world (Nicolau 2019). In this way, Spotify sought to associate music consumption in Latin America with the notion of streaming rather than "piracy" or "illegality."

Compared to the case of Netflix, relatively few researchers have analyzed the evolution and implications of Spotify's presence in Latin America. Pre-dictably, most research has been conducted in Brazil and Mexico, the two countries in the region that have the most Spotify users. This research has focused on understanding cultural patterns and preferences revealed by the most popular music streamed on the platform (de Melo, Machado, and de Carvalho 2020; Demont-Heinrich 2019; Pérez-Verdejo et al. 2020), or on how users make sense of Spotify's most iconic features, namely, playlists and algorithmic recommendations (Moschetta and Vieira 2018; Siles et al. 2019b; 2020; Woyciekowski and Borba 2020). What is available has been discussed mostly in news and cultural media publications. In these outlets, Spotify representatives have emphasized the notion that the platform has helped Latin American artists reach their full potential by achieving success beyond the region. Mia Nygren offered Spotify's (2019a) version of this phenomenon:

> In Latin America, we've helped restore a rapidly shrinking industry to growth. [. . .] The industry is healthier than it has been in years. Latin America propels local acts onto the global charts, and the rest is history. Music in the Spanish language has never had such a big audience thanks to ease of distribution.

In this account, Spotify's contributions to Latin America are twofold. First, it helped "restore" the music industry to "health" (presumably alluding to the "ill" of piracy). Second, it offered specific distribution tools for Latin American music to reach world audiences. One such tool was the playlist. As Eriksson and colleagues (2019) show, Spotify strengthened the playlist's status as the center of the streaming universe as part of its curatorial turn. The preferred example to illustrate the importance of Spotify's playlists for Latin American music has been the popularization of reggaeton. According to Spotify representatives, the platform's 2015 *Baila Reggaeton* playlist was a turning point in the "unstoppable rise" of the genre (Resto-Montero 2016). Rocio Guerrero, Spotify's then head of Latin content management and the person who created the playlist, noted in 2016: "The reggaeton audience today is bigger than any other Latin genre. The reality is this playlist is being played in Singapore, London and all over the world. There are Latin people all over the world, but it's not just Latin people listening" (Resto-Montero 2016).

Since its creation, the *Baila Reggaeton* playlist has become one of the most streamed on Spotify (Spotify 2018). According to some cultural commentators, it is the genre's "kingmaker" (Resto-Montero 2016). Guerrero further suggested that to reach global success, reggaeton had to become "platform ready"—adapting Gillespie's (2014, 168) expression "algorithm ready"—it had to transform itself to suit Spotify's definition of successful tones, sounds, and lyrics (Resto-Montero 2016).

This account thus emphasizes Spotify's importance in helping Latinx music and artists transcend the boundaries of regional success. But the other side of the coin warrants some consideration, namely, how the platform has benefited from music created in the region and how such cultural phenomena as reggaeton have fueled the use of Spotify both in Latin America and abroad.

By the turn of the decade, Spotify emphasized again the "streaming muscle" of Latin America by showing the growth of podcasts in the region (Spotify 2019a). It also claimed to honor the "Streaming Capital of the World" by holding the first version of the Spotify Awards in Mexico City, in early 2020 (Merino and Huston-Crespo 2020).

TIKTOK

TikTok started becoming popular in Latin America in 2018, a few months after ByteDance launched it outside of China and merged it with Musical.ly, a lip-synching video-creation app they had acquired. By the end of that year, Mexico and Brazil were respectively the fifth and sixth countries in the world with the most TikTok downloads (Argintzona 2020). As part of its efforts to expand geographically, in that same year, TikTok also assigned country managers in Brazil and Mexico to coordinate issues related to the use of the app and the creation of content in Latin America. According to Latinobarómetro (2018), 89 percent of Latin America's adult population owned a cell phone in 2018. GSMA (2020) estimated that, by 2020, 57 percent of the region's population had access to mobile Internet, a significant increase compared to 5 years before.

A considerable number of news articles about TikTok were published in Latin America throughout 2019. These articles offer a window on the evolution of the cultural meaning of the app. In these publications, the writers emphasized the idea of a technological revolution and invited users in the region to become part of it. Mexico's *El Universal* thus made sure to let readers know why TikTok would surely become their "next favorite social media platform" (Salcedo 2019). Argentina's Infobae (2019) guaranteed readers they would be kept updated on everything they "needed to know about TikTok, the app that has become a worldwide phenomenon." Media organizations also often discussed TikTok's "accelerated and exponential growth" (Salcedo 2019) and frequently mentioned that TikTok had been one of the most downloaded apps both for Android and iOS during 2019, with more than 1.5 billion downloads worldwide (K. E. Anderson 2020; Perretto 2019). Repeating these statistics suggested that downloading the app was almost inevitable. News outlets presented TikTok as "the next big thing" in the evolution of social media, and they spoke to the role of the app's country managers in emphasizing what made it supposedly "unique." They stressed the importance of algorithmic personalization and the ease provided by the app to create content. Said Noel Nuez, responsible for the app's development in South America:

> Videos are recommended based on the preferences of each user to ensure a personalized experience. [. . .] Another factor that differentiates us is that our platform has powerful and simple tools for video editing. Whereas YouTube is a showcase for a minority to create content, TikTok is a creativity laboratory where anyone can produce content. (Prieto 2019, my translation).

The increasing number of news articles about TikTok also revealed the role of country managers in promoting the creation of local content as a tool for "personalizing" the app's experience. A TikTok representative in Mexico thus indicated that this country stood out as the heaviest user in the region (Salcedo 2019). He also attributed the source of this success to what he thought was an intrinsic propensity of Mexicans for using social media, and he made sure to differentiate the preferences of people in such cities as Guadalajara, Mexico City, and Monterrey. Overall, this discourse posited that Mexico occupied a distinct position in the TikTok "revolution."

The notion that Latin Americans are more prone to use social media than other people in the world has been a common trope in the media. In this view, it was now TikTok's turn to "benefit [. . .] from that love of online socializing" that was naturally attributed to Latin American users (Ceurvels 2020). A TikTok representative adapted this discourse to the case of Brazil:

> We love Brazilian users because they are some of the funniest and most creative of all. Since it is the largest population in Latin America, TikTok has a huge potential to allow more Brazilians to show their talent and creativity as well as participate in many exciting challenges in the app. Our focus is growth by engaging our creators, introducing more diverse content, and continuing to strengthen our local presence. (Perretto 2019)

Again, TikTok representatives emphasized how the app could offer a means to potentiate supposedly intrinsic qualities of Latin Americans—such as humor, creativity, and talent—particularly in the case of teenagers. As Mexico's country manager put it, TikTok hoped to "inspire" Latin Americans to participate in its "global community" of users by creating content on the platform with local singularities (Salcedo 2019). Several sources reported that, by the end of 2019, TikTok had almost 20 million users in Mexico and 18.4 million users in Brazil (Argintzona 2020; iLifebelt 2020). In addition to TikTok's success in these two countries, it was being downloaded in great numbers in Argentina, Colombia, and Ecuador, as well as in Central American countries. A 2020 report estimated that 10 percent of TikTok users were located in Latin America, an identical number to that of Europe (10 percent) and similar to the United States (12 percent) (Iqbal 2020).

For most of 2020, the media focused on "success" stories that portrayed the cases of "TikTokers" with the biggest number of followers on the platform. These stories tended to work as practical invitations to become an "influencer" in Latin America. In parallel, some commentators also reflected on Donald Trump's decision to block downloads of TikTok in the United States and the eventual implications for Latin America (Malaspina 2020). Underpinning these reflections was the premise that, above everything else, TikTok had become a political phenomenon (Kleina 2020; Moreno 2020).

WHY COSTA RICA?

As the previous section established, Latin America holds a strategic position in both the present and the future of the worldwide digital economy. More specifically, this book focuses on Costa Rica. This country has some characteristics that make it an ideal study case in Latin America. The focus on Costa Rica allows a discussion of the findings from one of Latin America's most digitized countries in terms of infrastructure, use, and culture. Costa Rica's telecommunications infrastructure and per capita number of social media users are comparable to the largest countries in Latin America, which helps capture broad tendencies in the use of platforms across the region (Latinobarómetro 2018; Siles 2020). However, compared to such countries as Argentina, Brazil, Colombia, and Mexico, Costa Rica's local cultural production scene (which includes television, film, and music) is significantly smaller. This is important in that it creates a relatively different cultural need or place for entertainment platforms, such as Netflix, Spotify, and TikTok. In this regard, Costa Rica is more typical of Central America and small countries in South America.

Before delving deeper into the book's chapters, I offer a brief account of some of the country's most important historical, cultural, and technological specificities and how these shape people's relationship with algorithmic platforms.

Since the abolition of its army in 1948, Costa Rica has consistently and strategically built its national identity around the idea of being a peaceful nation in the middle of a typically turbulent region (Sandoval 2002). The construction of the country's public image has also rested on the notion of being the oldest democracy in Latin America. More broadly, Costa Rica is usually associated with some of its most important economic sources of income: quality coffee in the previous decades, and more recently, tourism, technology, and sustainable development (Ferreira and Harrison 2012). And after 2011, when it topped the Happy Planet Index for the first time, Costa Ricans also like to refer to the country as "the happiest nation in the world." Mitchell and Pentzer (2008, 227) summarize this mythical view of the country:

> Costa Rica generally thinks of itself as an inherently democratic, peaceful, and ethnically homogeneous society. [. . .] This is, then, an image of Costa Rica as *exceptional*, not at all like its Central American neighbors. Catholicism is another part of the picture; it is still the official state religion. [. . .] A final element is the generalized belief in Costa Rica as a middle-class society.

This dominant sense of nationhood is predicated on a particular historical interpretation of the colonial period, forged during the eighteenth and nineteenth centuries. According to this interpretation, the deprivation and poverty that characterized the country during colonial times brought about equality among its inhabitants (Sandoval 2002). It was also during the nineteenth century that the "the tradition of small landowning in the Central Valley led to a kind of natural 'rural democracy' and a distinctive national psychology inclined toward compromise rather than conflict" (Mitchell and Pentzer 2008, 228).

In this book, I examine an underexplored dimension of the country's daily life, namely, its relationship with technology and digital media. Over the past three decades, technology has been established as a key element in Costa Rica's national identity. Costa Rica was the first Central American country that connected to the Internet, and the country played a catalyzing role in the Internet's development in this region (Siles 2008; 2020). After Intel inaugurated one of its largest manufacturing plants there in 1998, the country's capital (San José) got the fleeting nickname of *San Jose South*, an "ambitious nod to San Jose, California, the heart of Silicon Valley," in the words of a *Wall Street Journal* reporter (Vogel 1998). This particular event galvanized a major shift in the country's national identity: becoming a "technology nation" was embraced as a way to dissociate from being a "banana republic" (Siles, Espinoza-Rojas, and Méndez 2016).

Since the 1990s, multinational companies have settled in Costa Rica in large part as a result of initiatives to attract foreign direct investment. Several factors account for this: the characteristics of the country (its relative political and social stability, ongoing economic liberalization processes, previous experience in the development of electronics, receptive attitudes toward foreign investment, quality of education, geographical location, and climate, among other traits); the negotiation tactics used by the government and some nongovernmental organizations; and the specific concessions given to multinational corporations (such as tax incentives, support in transport infrastructure, and agreements with universities to strengthen vocational training projects) (Larraín, López-Calva, and Rodríguez-Clare 2001; Nelson 2009; Spar 1998).

As a result, the country has experienced exponential growth in the production of such technologies as software, video games, e-learning, digital animation, and the creation of cell phone applications (Nicholson and Sahay 2009; Paus 2005). By 2015, a study classified San José as the "most competitive city in the world" for investment in the production of communication technologies (Conway et al. 2015). Today, the country is the largest per capita exporter of high technology in Latin America (M. E. Porter and Ketelhohn 2002; Soto Morales 2014). More recently, as neoliberal ideas associated with the notion of startup entrepreneurship have spread, some commentators have expressed a desire to turn the nod to San Jose (California) into a more explicit message: transforming Costa Rica into the "Latin American Silicon Valley" (Ciravegna 2012; Siles, Espinoza-Rojas, and Méndez 2016).

A similar scenario characterizes the use of media technologies in the country. In 2020, 96.3 percent of Costa Rica's population said they owned at least one cell phone (96.4 percent in urban areas and 96.0 percent in rural areas) (INEC 2020). According to Costa Rica's National Household Survey, the most reliable source on the matter, almost 85 percent of households in the country had some form of Internet access in 2020, mostly through broadband infrastructure (INEC 2020). The country also leads Latin America in the use of numerous social media platforms, including Facebook and WhatsApp (Latinobarómetro 2018). A report from the United Nations Economic Commission for Latin America and the Caribbean (ECLAC) characterized the spread of mobile Internet access in Costa Rica as a "model" in Latin America (CEPAL 2016). This report also showed that the country has the smallest gap in the use of cell phones between urban and rural zones in the region, a factor that helps mitigate the significance of the digital divide in Latin America (which characterizes some of the largest countries in the region). At a more qualitative level, technology plays a central role in shaping the conditions of daily life in the country. It has become the center of people's aspirations and the object of their admiration (Cuevas Molina 2002).

How the tension between the local and the global is experienced in the country offers another crucial element for understanding Costa Ricans' relationship with technology (including algorithms). As the next chapters demonstrate, the use of algorithmic platforms enacts a tension between reproducing traditional myths and values in Costa Rican society and participating in a global consumer society shaped by ideals from the global north, most notably the United States. On one hand, mythical views of the country (such as a supposedly intrinsic proclivity for peace and the desire to maintain the status quo so as not to disturb the social order) provide people with resources to enact algorithms in specific ways. As I will show, people tend to "force" algorithms to comply with local rules and traditions of social interaction. On the other hand, Costa Rica has historically been a fertile ground for ideas, values, ideas, attitudes, and products from the United States (Cuevas Molina 2003). As Harvey-Kattou (2019, 5) puts it, a "facet of Costa Rican identity which lends itself to the rhetoric of [. . .] exceptionalism is its close relationship with the USA." In the case of algorithms, Costa Ricans extend this ideal of cultural proximity by reproducing certain practices and consumption patterns that characterize the United States. For this reason, I argue that the similarities in how Americans and Costa Ricans enact algorithms should be considered a finding in itself. In other words, these similarities are a product of how Costa Ricans channel their own cultural aspirations through technologies such as algorithms.

RESEARCH DESIGN

In this book, I analyze how ordinary technology users in Costa Rica relate to three algorithmic platforms in the specific domains of culture and entertainment. This analytical preference shaped both the findings and the claims made in the book. Unlike other kinds of platforms, users pay for the services of Netflix and Spotify or download TikTok to receive content that connects them to other people both locally and globally. Thus, the theoretical and empirical approaches developed in the book can provide numerous insights for studying other entertainment platforms. (Further research could shed more light on how this approach could be used to study algorithmic devices in systems of governance or finance.)

This book draws on a wealth of empirical evidence collected over a 5-year period in Costa Rica. I sought to operationalize a research design with "ethnographic sensibility"—borrowing Star's expression (1999, 383)—devised "as a practice for producing and participating in plural enactments of algorithms" (Seaver 2017, 10). Although this type of methodological approach is ideal for "map[ping] different values for data evoked in different discourses [. . .] and contexts" (Fiore-Gartland and Neff 2015, 1471), it is less well suited for statistical analysis or generalization. The analysis also centered on people's accounts of their practices rather than a direct observation of their daily life activities. The focus on cultural dynamics also meant that my research was not designed to single out how specific variables (such as gender, age, or class) shaped awareness of algorithms. Some scholars have engaged in such work in an exemplary way (Gran, Booth, and Bucher 2021; Hargittai et al. 2020). However, as the next chapters demonstrate, I remained attentive to the significance of intersectional markers of identity in people's relationship with platforms. (The book's appendix offers a more detailed discussion of how the research for each chapter was conducted. It explains the specific methodological procedures employed to sample participants for each method, collect the information, and analyze the data through inductive and abductive strategies.)

I employed three main research methods. First, between 2017 and 2019, I conducted 110 interviews with users of Netflix and Spotify in Costa

Rica. I began by sharing a call for participation on social media and selected individuals with different profiles among those who responded. To increase diversity in my sample, I asked interviewees for additional suggestions of people with different backgrounds. I selected both "heavy" users and more casual users for interviews. The final groups of interviewees included mostly educated people with a diversity of professional backgrounds who were at different stages in their careers. I conducted all interviews in person. These interviews lasted an average of 43 minutes. Most conversations took place in the School of Communication at Universidad de Costa Rica in San José.

During the interviews, I asked participants about their practices when using these platforms. With their approval, I also asked them to open their account on a computer and project content on a screen. I used these projections to ask informants about specific content available in their accounts, technical configurations, and algorithmic recommendations. Although few interviewees used the term "algorithm," many had thought about how recommendations worked and felt they understood the logic of the recommendation process. We discussed at length these convictions and certainties about the operation of these recommendations. To foster data-source triangulation, I also captured screenshots from users' Netflix and Spotify accounts. I compared descriptions of how algorithmic platforms work according to users with those provided by the Netflix, Spotify, and TikTok representatives in the mainstream media and official outlets.

In addition to interviews, I conducted twelve focus groups with fifty-seven participants located in this country. (These participants differed from those who were interviewed.) I recruited these individuals through a call for participation that was available on social media profiles associated with the university where the research was conducted (Universidad de Costa Rica). Potential participants were asked to fill out an online questionnaire, which allowed me to select individuals with different sociodemographic characteristics. The first four focus groups (with twenty-two participants) centered on the use of Spotify and were conducted between August and October 2019. Participants were between 18 and 62 years old.

The second set of focus groups (eight in total) were conducted with thirty-five Costa Rican TikTok users (both people who created videos and those who exclusively watched them) between June and July 2020. Participants in the TikTok focus groups were between 18 and 56 years old. Since this part of the research was conducted during the COVID-19 pandemic, focus groups about TikTok use were conducted on Zoom. Focus groups were ideal for exploring how people developed their ideas as they shared them with others. Accordingly, I examined the dialogs, discussions, and collective construction of ideas about algorithms that unfolded during these sessions (Cyr 2016).

Finally, I carried out a third research method, namely, rich pictures. A building block of the so-called "soft systems methodology" (Checkland 1981), rich pictures consist of diagrams or drawings made by individuals to graphically represent a specific phenomenon. I turned to this technique to have access to people's unstated and taken-for-granted knowledge of algorithms. I provided participants in focus groups with blank sheets of paper and a set of pens; I then asked them to individually draw how they thought algorithmic platforms operated and offered them recommendations. Participants in focus groups explained their own pictures and discussed aspects of other participants' drawings. I then used the Bell and Morse (2013) guide to analyze pictures and thus examined descriptive features and structures (such as the use of colors, shapes, thickness, relationships, and arrangements).

OVERVIEW OF THE BOOK

Rather than focusing on one specific platform, each chapter discusses different aspects of users' experiences with Netflix, Spotify, and TikTok in Costa Rica. In this way, I develop larger cultural dynamics in each chapter by integrating people's relationships with different platforms.

Chapter 2 focuses on issues of personalization. It challenges dominant accounts of personalization that focus on users' desire to receive "personal" recommendations only. Instead, I redefine personalization as the establishment of communication relationships between users and algorithms. In short, I view personalization as a communication process through and through. To make this case, the chapter considers three dynamics of personalization. First, users experience algorithms as a form of *interpellation*. I focus on the case of Netflix users in Costa Rica to show how they perceive the platform's algorithms and interface as mechanisms that "hail" them in personal ways. Second, personalization also implies the *personification* of platforms, that is, the attribution of person-like characteristics that shape users' understanding of algorithms. I examine how users in the country conceive of Spotify as a social being that intermediates their personal relationships.

Third, interpellation and personification pave the way for establishing particular kinds of *relationships* with algorithms. By examining the case of TikTok users, I argue that the relationship with algorithms is neither stable nor linear, as has been suggested in the literature. Instead, it undergoes a set of "passages" through which attachment to the platform emerges and is performed. Together, these three dynamics show how personalization is a cultural achievement that requires weaving relationships with algorithms.

Chapter 3 offers a much more nuanced account of algorithms' influence in society than what has been offered in the scholarly literature on datafication. I invert the dominant focus on how algorithms shape users' choices by examining instead how users integrate algorithmic recommendations into the structure of their daily lives. The notion of integration emphasizes the practices of individuals who variously combine cultural elements as they ponder what recommendations they want to follow. Integration is the process of enacting a repertoire of cultural resources to respond to the situations people face in their daily lives. This chapter discusses three types of integration work. First, I examine how Costa Rican Netflix users combine a set of cultural *sources* as they consider what content they want to watch. Second, I analyze how Spotify users interpret algorithms as opportunities to integrate certain *capacities* into their lives. Finally, considering the case of TikTok, I show that users' understanding of algorithms stems from practices with infrastructures and technologies other than this platform. Algorithms thus work as *relational categories* that users employ to integrate their experiences with other platforms into their relationship with TikTok (and, vice versa, use their relationship with TikTok to assess other platforms). By discussing how users integrate cultural sources, capacities, and relations, I provide an analysis of why people follow algorithmic recommendations (or why they do not follow them) that emphasizes the significance of culture. I thus argue that algorithms do not work entirely alone, nor do they act as the only determinant of cultural consumption.

The mutual domestication between users and algorithms also takes shape through patterned practices that occur at specific times and places. Chapter 4 theorizes these practices as rituals. Analyzing rituals brings power to the forefront of the analysis. Drawing on Nick Couldry's work, in this chapter, I discuss a particular kind of myth that is enacted and reproduced through the ritual use of algorithmic platforms. I call this the "myth of the platformed center," the belief that algorithmic platforms are the center of people's social world. I argue that rituals around Netflix, Spotify, and Tik-Tok normalize the idea that users have to cultivate practices, moods, and emotions, and that algorithmic platforms are an obligatory intermediary in this process. I first consider the types of rituals that characterize the domestication of Netflix and its algorithms. Netflix normalizes recommendations by suggesting that they are the product of the users' own rituals (these recommendations are offered to users "because they watched" content previously). The chapter then argues that rituals infuse algorithms with affect. I develop this notion by discussing how users have embraced the creation of their own playlists on Spotify as a means to produce, capture, and explore moods and emotions. Finally, the chapter looks at the case of TikTok users who turn to this app to deal with boredom. I examine boredom as a moral emotion that users feel obligated to resolve. To this end, users develop specific kinds of rituals through which they act out the centrality of TikTok in their lives. In short, this chapter argues that the power of algorithmic platforms relies on the unceasing reproduction of rituals.

According to Silverstone (1994), conversion is the process of reconnecting with the public world through technology or its contents. Conversion turns people's personal relationships with algorithmic platforms into a public issue. It involves displaying, sharing, and discussing recommendations with others. It requires certain skills, practices, and resources. Chapter 5 examines these issues by further situating conversion within larger "regimes of publicness," that is, the articulation of mutually defining conceptions of self, publicness, and technology. I argue that conversion articulates the self to others through algorithms and connects the self to algorithms through others. The chapter begins by discussing conversion dynamics on Netflix and how users transform their knowledge of the platform into an opportunity to influence others in their decision process. I show how people use the platform and its algorithms as an opportunity to define their status, sense of belonging, and affect as part of networks of interpersonal relationships. Then the chapter draws on Lauren Berlant's (2008) notion of "intimate public" to analyze how playlists created by users form the basis of collective experiences that serve Spotify's political-economic project. Algorithms thus become opportunities to enable and bring about this kind of publicness. The chapter concludes with a discussion of the networks of exchanges that take place on TikTok. In this case, conversion is best understood as a collective technique through which a sense of "close friendship" is enacted. By examining these conversions (in the plural), the chapter emphasizes users' capacity to inhabit multiple registers of publicness through and with algorithms.

Despite their appreciation for algorithmic platforms, users in Costa Rica also criticize and challenge certain aspects of how they operate. This permeates the relationship with recommendation algorithms and, thus, ends up shaping the mutual domestication process. Chapter 6 refers to these critical notions as "resistance." Most work on resistance in critical data studies has focused on explicit attempts to change the pillars of datafication. Instead, this chapter focuses on more subtle forms of "infrapolitics," or low-profile forms of resistance. Infrapolitics are resistance practices that operate at the interstices of everyday life but that lack political articulation. These practices express people's sense of autonomy within the datafication system rather than attempt to change it. First, I consider how users reacted against what they perceived as biases in Netflix's recommendations and how they claimed their senses of identity against these biases. Netflix users in Costa Rica rejected and resisted algorithms when it became clear that platforms didn't consider them as people but rather as consumer "profiles." Second, I discuss resistance to Spotify's attempt to "configure" users into paying customers and the "obligation" to make algorithmic recommendations the center of their music consumption practices. Finally, I analyze how users made TikTok's political project obvious through a twofold dynamic: the need to delete what TikTok promotes while promoting what TikTok tends to delete. Drawing on these examples, the chapter concludes by proposing that the notion of hegemony offers a more useful framework than alternatives (such as data colonialism) to make sense of processes where domination through datafication and resistance intersect. This is because culture is a domain of hegemonic dispute.

Chapter 7, the final chapter, discusses the implications of the evidence presented in the book for theorizing about the relationship between users and algorithms. I begin by summarizing the main claims made in the book and situating them against the background of prevalent technological determinism. I then develop the notion of mutual domestication. To this end, I draw on Silverstone's discussion of domestication as a cyclical process, Giddens' theory of structuration, and recent work in science and technology studies on multiplicity. I build on this scholarship to discuss two related processes: the place of algorithms *in* Costa Rican culture and how algorithms can be considered *as* Costa Rican culture. Through the notion of mutual domestication, this chapter argues that the study of algorithms in and algorithms as culture are simultaneous occurrences: algorithms are designed to adjust to people's practices in order to domesticate users, but users enact algorithmic recommendations in their daily lives. Mutual domestication implies that datafication is not a unilateral process. Accordingly, understanding this process requires broadening the way in which agency has been framed in current discussions of datafication. I argue that our understanding of agency must account for people's capacity to simultaneously follow and resist algorithms in their lives. This chapter elaborates on what this argument means for the study of the relationship between users, algorithms, and culture.

PERSONALIZATION

Personalization has been key in the study of datafication. For example, Van Dijck, Poell, and De Waal (2018) considered it to be one of the main operational principles of platformization. Most researchers have focused on the mechanisms of personalization such as "profiling," that is, the transformation of users into "profiles" or "data doubles" that, based on the constant surveillance of their practices, "enable corporate actors to refine their influence on users" (Couldry and Mejias 2019, 131). According to these researchers, "profiles" work as models for user behavior or "discursive parameters of who [users] can and cannot be" (Cheney-Lippold 2017, 48). Scholars have insisted that, through mechanisms such as "profiling," algorithmic personalization "imbues [technological systems] with the power to co-constitute users' experience, identity, and selfhood in a performative sense" (Kant 2020, 12). This form of algorithmic power is premised on both the extractive practices of technology companies and users' willingness to share their own data in order to receive "personally relevant" information. Researchers have applied this view of personalization to the study of news (Monzer et al. 2020; Powers 2017), music streaming (Prey 2018), audiovisual streaming (Novak 2016), marketing (Kotras 2020), and contemporary sociality (Lury and Day 2019), among many other areas.

As Cohn (2019) and Kant (2020) have shown, personalization has a long cultural and technological history. Both authors traced the emergence of personalization to technological advancements in the mid-1990s, which acquired a new scale of development throughout that decade as pieces of code (such as "cookies") and targeted advertising rose to prominence in the nascent Web economy. The development of algorithms helped cement recommendations into an industry in itself (Cohn 2019). For Kant, the centrality of algorithms in the personalization process implies a shift in responsibilities: "it is not the user who is responsible for the implementation of personalization, but algorithmic protocols that are designed to automatically render an object personal in the user's stead" (Kant 2020, 30). In this view, personalization is something that platform companies provide.

In this chapter, I provide a different way of conceptualizing personalization and its significance for datafication. Authors such as Kant speak of instances when users "encounter" personalization that is offered to them by algorithmic procedures. Instead, I argue that personalization is not something that is "encountered" but rather is built or established between users and algorithms through communication dynamics. Accordingly, personalization stems from more than a desire to receive content with personal relevance. Personalization, I contend, is in essence a communication process; it is the establishment of a *personal* relationship with platforms and algorithms. This alternative view posits personalization as a cultural process in that it makes explicit (rather than takes for granted) the meaning of personal communication and relationships in particular cultural contexts.

In what follows, I operationalize this approach by discussing three specific personalization dynamics derived from studies of how people interacted with algorithmic platforms. First, personalization refers to how users in Costa Rica experience algorithms as a form of personal *interpellation*. I focus on the case of Netflix users to show how they perceive algorithms as mechanisms that "hail" them, that is, as a form of personal interpellation. I consider how Netflix both reproduces and extends traditional mechanisms of personalization that have prevailed in previous media. Althusser's account of interpellation helps further understanding of a second dynamic of personalization, namely, the *personification* of platforms and algorithms. I use Spotify to show how users attribute human-like characteristics to platforms and algorithms. By so doing, they envision technology as a social intermediary of personal relationships and thus normalize surveillance practices as a necessary condition for receiving relevant recommendations. Finally, personalization culminates in the creation of a specific kind of *relationship* with algorithms. I focus on the case of TikTok users who, building on notions of interpellation and personification, establish a relationship with this app's algorithms that shifts continuously as it undergoes a series of "passages." I turn to actor-network theory to argue that the user–algorithm relationship must be thought of as constantly "in the making" rather than stable and fixed.

ALGORITHMIC INTERPELLATION ON NETFLIX

Rosaura is a 25-year-old public health student at the largest public university in Costa Rica. She is a devoted fan of so-called "romantic comedies." In her words, these movies allow her "to disconnect from everything" and make her "feel well." She made sure to clarify: "it's not like I'm that way in real life, but they [romantic comedies] do evoke nice things in me when I watch them, like some kind of peace." To watch these movies and gain this sort of "peace," Rosaura always turns to Netflix. Not only because it has some of her preferred movies from the 1990s but also because she can constantly discover new series and films. The users of algorithmic platforms that I interviewed employed expressions that are consistent with the belief that someone was hailing them or talking specifically to them through recommendations. When asked how she assessed algorithmic recommendations on Netflix and made decisions about which content to watch, Rosaura was straightforward: "they have to *call* me" (emphasis added). Rosaura thus expressed how she felt that Netflix was speaking directly to her by recommending certain kinds of content. Other people used expressions such as "this recommendation summons me" (*me convoca*) to explain why they had found a recommendation appropriate.

The belief that users are being addressed in a personal manner through algorithms is a first key to understanding personalization. I theorize this process as *algorithmic interpellation*, that is, the work embedded in algorithms to convince users such as Rosaura that Netflix is speaking directly to them, "hailing" them in particular ways. Rather than looking only at the particularities of the content that is recommended to audiences, as has been a dominant approach in scholarly literature, in this section, I first examine how a group of Costa Rican women made sense of the algorithms that drew their attention to romantic content (or how the content is recommended). The focus is on algorithmic recommendations associated with romantic love because of its pervasiveness in public culture and its significance in revealing the reproduction of capitalist patriarchal structures through interpellation (Radway 1984; Swidler 2001a). As Butler (2016, 18) put it, gender assignment has "an interpellating force." Most studies on similar issues have concentrated on the internal biases of algorithms and the affordances of recommendation platforms themselves (Eriksson and Johansson 2017; Werner 2020). I supplement these approaches with an analysis of what women actually do in practice when they encounter biased recommendations. By looking at both how a group of women related to algorithms and the mechanisms that Netflix employed to make recommendations appealing to them, I espouse a view of gender and technology as mutually constitutive (Wajcman 2004).

INTERPELLATION AND ALGORITHMS

Althusser (2014) argued that individuals become subjects through the way they are ideologically addressed or "interpellated." For Althusser, ideology works by hailing individuals. Interpellation, he argued, "can be imagined along the lines of the most commonplace, everyday hailing [. . .]: 'Hey, you there!'" (Althusser 2014, 190). When individuals recognize themselves as the addressees of certain discourses, they accept the subject position that is offered to them. Fiske (1992) built on Althusser's ideas to consider the media's capacity to interpellate audiences. Although his ideas are more theoretical than empirically derived, he summarized this process in a useful manner:

> In communicating with people, our first job is to "hail" them, almost as if hailing a cab. To answer, they have to recognize that it is to them, and not to someone else, that we are talking. This recognition derives from signs, carried in our language, of whom we think they are [. . .] In responding to our hail, the addressees recognize the social position our language has constructed, and if their response is cooperative, they adopt this same position. (Fiske 1992, 217)

For Althusser (2014), this process also supposes the existence of a unique and central Subject, which interpellates all individuals as subjects. He used Christian religious ideology to illustrate this point. In this account, God is the Subject and God's people are the "Subject's interlocutors, those He has hailed: His mirrors, His reflections" (Althusser 2014, 196). Interpellation thus leads to the establishment of power formations through which roles, norms, and values are assigned and reproduced.

This interpellation framework can fruitfully be applied to the case of algorithms. As Cohn (2019) argues, algorithms interpellate users by offering them certain content to which they can relate but also by demanding that users think of themselves as subjects who could desire these suggestions. For Cohn, algorithmic interpellation functions when users agree to think in the terms offered through the interface of a platform. Drawing on a phrase from Lev Manovich, he thus defines interpellation as mistaking for one's own the structure of someone else's mind embodied in a platform. Cohn aptly describes how interpellation through algorithms operates as a dynamic of mutual domestication: "When we try to make our recommendations 'better' [. . .] we feel as if we are mastering the program, but all the while, the program is mastering us" (Cohn 2019, 8). Moreover, Cohn notes, interpellation works in algorithmic platforms by establishing subjectivation through social comparison. In other words, platforms invite people to think that algorithmic recommendations are offered to them because such recommendations represent or belong to certain kinds of people ("users like you").

For the most part, studies have focused on Netflix's role as a content producer. Whereas some have celebrated the variety of content available on the platform that was written, produced, and features women (Bucciferro 2019), others have shown important continuities between Netflix's "original" productions and gendered norms that "reaffirm hegemonic discourses through [shows that] disguise [themselves] as ironic and progressive social critique" (Rajiva and Patrick 2021, 283). But in addition to this role (which connects the platform to television studies), Netflix is also a particular technological infrastructure (which links it to technology studies). Netflix is both television *and* new media; it is a "texto-material assemblage" (Siles and Boczkowski 2012). As Lobato (2019, 43) puts it,

> Netflix is a shape-shifter: it combines elements of diverse media technologies and institutions. [. . .] In its dealings with government, Netflix claims to be a digital media service—certainly not television, which would attract unwelcome regulation. Yet, in its public relations, Netflix constantly refers to television, because of its familiarity to consumers.

Thus, for Netflix, interpellation operates at two simultaneous levels: the media texts or discourses that present users with certain subject positions; and a socio-technical system that works to interpellate specific individuals. Studies have focused on the former dimension, leaving the latter with comparatively less attention.

Algorithmic recommendations offer a key opportunity to examine how Netflix interpellates users (and women in particular) as a sociotechnical system. Seen in this way, Netflix's algorithmic recommendations are a condition of possibility for the interpellation of individuals and their transformation into ideal subjects (or specific kinds of consumers) through personalization.

"WHO'S WATCHING?"

Netflix has crafted a public identity as a company that specializes in personalization. As Netflix's Chief Content Officer, Ted Sarandos, put it, "our brand is really about personalization" (cited in Rohit 2014). The work of personalization is accomplished in a variety of ways: "at Netflix, we embrace personalization and algorithmically adapt many aspects of our member experience, including the rows we select for the homepage, the titles we select for those rows, the galleries we display, the messages we send, and so forth" (Chandrashekar et al. 2017). The company has emphasized the idea that it does not offer a single product for all users, but rather "over a 100 million different products with one for each of our members with personalized recommendations and personalized visuals" (Chandrashekar et al. 2017).

Accordingly, Netflix's interface is designed to make users feel they are receiving content that has been prepared exclusively for them. Althusser's classic interpellation formula, "Hey, you there!" is enacted through the mandatory process of creating a profile on the platform. Thus, the first question that Netflix asks users is, "Who's watching?" Before being able to navigate and watch content, users need to select a profile (see figure 2.1). This invites users to recognize themselves as interpellated subjects. All users I interviewed had their own profile, except for one person (who shared it with her husband). Users tended to name these profiles after themselves and to choose specific avatars among Netflix's pre-selected image catalog.

Practically all the interviewees who selected images (rather than using the default smiling face) chose an image of a woman. Fernanda, the Communications Director of a transnational enterprise, selected the avatar shown

Figure 2.1
Fernanda's *morena* female avatar (fourth from left), taken from Netflix's "Who's Watching" user profile page.

in figure 2.1 (fourth from left). She explained: "I don't know if there is an equal number of male and female [avatars], but this is the only female *morena* (brown skin) avatar there is." Like Fernanda, the goal of users who chose an image was to find one that best captured a defining feature of themselves (whether a physical attribute or a personality trait). Fernanda thus challenged Netflix's default setting by choosing the only female character available in the catalog that she related to. This practice thus reinforced the notion that the platform reflects the personality and self of users.

Answering the question "Who is watching?" then paves the way for other interpellation mechanisms on the platform's interface. The content is displayed under sections such as "Continue watching for *you*," "Recommended for *you*," "Top Picks for *you*," and "Because *you* watched" (emphasis added). Using the profile on the platform works to confirm the belief that personalization is Netflix's most distinctive interpellation feature. Natalia, a 20-year-old college student, noted: "I once used my mother's profile by mistake and everything looked different. I said, 'These are not *my* recommendations.' I realized then how personalized *my* algorithm is" (emphasis added). In this account, personalization also generated a sense of possession. Users reached similar conclusions when they compared Netflix to other platforms, such as HBO's streaming service. Whereas these platforms emphasized categories such as "Most viewed," users considered that Netflix emphasized recommendations "for you." They found the latter much more appealing. In this way, users defined content for *everybody* as the opposite of suggestions made exclusively for *them*.

As interviewees used Netflix more frequently, they intensified efforts to control aspects of their profiles. To make sense of this process, users typically employed the metaphor of "contamination": content that was chosen by another person polluted their own recommendations. Characterizing the situation in terms of contamination (i.e., a problem) invites a concrete response from users (i.e., a solution): actions to clean up any disorder, including the creation of more profiles or using third-party accounts to watch content about which users are unsure and think could contaminate their profiles.

People indicated they regularly provide the platform with feedback in order to improve recommendations. Many said they regularly used the "Thumbs Up" and "Thumbs Down" features to rate a movie or series they have watched. Natalia, the college student, noted that she frequently searched for movies on Netflix that she had watched on other venues with the specific purpose of "liking" it, so that Netflix could recommend to her what she thought was similar content. Recommendations were thus viewed as a result of the time "invested" (as one person described it) in providing such feedback.

Genres have traditionally played a key role in interpellation processes (Fiske 1992). For example, as noted in chapter 1, Martín-Barbero (1993) showed the centrality of *telenovelas* in interpellating popular classes in Latin America. In this perspective, melodramas historically have been an important genre through which people in Latin America have recognized themselves. Moreover, scholars have argued that genres function as "interpretive contracts" between producers and audiences through which certain expectations and subject positions are defined around the meaning of media texts (Livingstone and Lunt 1994). These interpretive contracts are not definitive but constantly contested and enacted in specific contexts.

The role of genres is problematized in Netflix's algorithmic interpellation. Netflix includes multiple features as part of its recommendations, including images, categories, names, content descriptions, recommendation percentages, and icons of thumbs up or down to reinforce the suggestion, among others. In short, Netflix does not offer a single recommendation but rather a bundle of features that tie algorithms and genre expectations together. Users can thus receive the same recommendation as part of different bundles. Figure 2.2 exemplifies how Netflix recommended its original

Figure 2.2
Four algorithmic bundles of *Anne with an "E,"* taken from the Netflix selection menu of four distinct users.

series, *Anne with an "E"* to four different women. I captured these screenshots during the interviews. Figure 2.2a was shown to Paula (a 39-year-old business administrator) under the category "TV Dramas"; figure 2.2b was recommended to Ema (a 39-year-old college instructor in human rights) in the category "Historical TV Shows"; figure 2.2c, which, unlike other images, shows both a female and a male character looking into one another's eyes, was suggested to Inés (a 25-year-old miscellaneous worker) as part of the category "Award Winning Shows"; and figure 2.2d appeared in Marcia's (a 19-year-old engineering student) profile under the category "TV Dramas Starring Women."

In a similar manner, I found Netflix's original series *Las Chicas del Cable* (*Cable Girls*) recommended in sixteen different categories in my interviewees' profiles, some of which were tied to romantic ideas ("Cheesy Shows in Spanish" or "Emotional European TV Dramas"), while others were not ("Historical TV Shows" or "Binge-worthy European TV Shows"). Netflix recommended the film *Always Be My Maybe* to Mariana, a 49-year-old labor lawyer, as part of two distinct categories: "Trending Now" and "Netflix Originals." These two categories, along with "Popular on Netflix" and "Recently Added," were common to all interviewees. The other algorithmic bundles varied from person to person. Some of the bundles that users received centered specifically on romantic content, but for the most part, series and movies associated with romantic love were included in all types of categories.

Traditional genres were key in how this group of women made sense of algorithmic bundles. In other words, they consistently applied generic rules to evaluate them. The most common traditional genre associated with the idea of romantic love that these women mentioned was the "romantic comedy" (or "rom-com"). This kind of content was appealing to many users, regardless of their age. Both younger and older users idealized romantic comedies produced in Hollywood in the 1990s and early 2000s (which they often referred to as "classics"). Carolina, a 21-year-old public relations specialist, argued that what distinguished these movies was how certain actresses portrayed specific female characters. In her words, "a romantic comedy is [always] the same. But I love the actresses. I know that I must have some kind of connection [with them]." In her account, movies such as *Bridget Jones's Diary* and any film that featured Jennifer Anniston or Sandra Bullock counted as a classic. These examples also reveal the significance of Hollywood products and ideals in the cultural imaginations of most interviewees. Algorithmic interpellation proved more successful for these women when it helped reaffirm the notion that Costa Rican women could identify, understand, and appreciate cultural products such as a "classic" Hollywood romantic comedy.

According to Radway (1984, 61), romantic love appeals to certain women because it creates a space in which an individual "can be entirely on her own, preoccupied with her personal needs, desires, and pleasures." Romantic content also works as a means to escape to the exotic or to what is different. This assessment applies neatly to the experience of the Costa Rican women I interviewed. They noted that romantic comedies offered them a narrative that interpellated them for various reasons. Fernanda expressed this idea eloquently:

> I believe that [Alfonso Cuarón's critically acclaimed movie] *Roma* has the same value that a romantic comedy has. I would even say that the romantic comedy touches more fibers than *Roma*. I think there is a much more universal language in romantic comedies than there is in an Oscar-winning auteur film.

Fernanda thus emphasized the affective value of romantic movies by suggesting that they can elicit strong feelings that connected her with other people. Other interviewees stressed hedonic reasons. When asked to explain her interest in romantic comedies, Viviana, a 20-year-old college student, blended the pleasures derived from movie watching and from eating as equal forms of escaping the burdens of daily life: "If I am very tired mentally then I look for some kind of snack, like popcorn or seeds, and then I sit on my bed, turn the TV on, and put on a 'chick flick.'" The fact that this "universal language" is repetitive provides them with a sense of control over such conditions. Speaking specifically about romantic comedies, Viviana continued:

> They're very predictable, you know what is going to happen. But it is still nice to have the feeling you get when character one says to character two that she loves him, and then comes a passionate kiss. I like that a lot, the happiness, the positive feeling it conveys.

My interviewees thus found opportunities for escape, pleasure, and control that resulted from their social position as middle-class women in Costa Rican society (cf Radway 1984). Netflix's interpellation worked successfully when these users felt that such bundles of symbolic features conveyed a meaningful recommendation that allowed them to fulfill their expectations of the feelings that a "rom-com" should evoke. Genres played a key role in this process. For users, personalization became most apparent when algorithms appropriately recommended the genres that most interpellated them. (In chapter 6, I also discuss instances of resistance to interpellation.)

In summary, interpellation on Netflix draws on the traditional centrality of genres to enable a sense of re-cognition in its users, in Martín-Barbero's sense (developed in chapter 1). It provides audiences with content that reproduces classic generic cues through which people can recognize themselves and their lives. But Netflix's content recommendations are also embedded in a larger sociotechnical system that reinforces this sense of interpellation. Key features on the platform (most notably its interface design and algorithmic bundles) strengthen the sense that users are being hailed by a very particular subject.

THE PERSONIFICATION OF SPOTIFY

A second dynamic of personalization consists of personifying platforms and algorithms by attributing person-like features to them. In this section, I analyze this dynamic by examining how users understood and interacted with Spotify and its recommendations. Personification rests on the interpellation process discussed previously. Interpreting algorithmic recommendations as a form of hailing leads to the notion of an interpellating subject. As users felt they were being addressed in a personal manner, they also imagined and attributed specific features to the subject that interpellated them. This also occurred in Netflix's and TikTok's cases. For example, users treated platforms such as Netflix as a central subject that spoke directly to them through recommendations. Elisa, the photographer, described her relationship with this subject in a typical manner: "Netflix and I know each other very well, because I do pay attention and heed (*le hago caso*). He then must think, 'Look, she heeded, so I will keep recommending things.'" In this account, Netflix is a male subject who not only communicates with his interlocutors but also rewards them for good behavior.

Reeves and Nass (1996, 5) famously referred to the process of personification as "the media equation," that is, the tendency to consider media technologies as "equal" to "real life." In this perspective, users attribute rules from interpersonal interaction rules to the media rather than treating the latter as mere tools, because users rely cognitively on the only thing that has been real for the human brain throughout its history: human social behavior and relations. Thus, Reeves and Nass (1996, 7) contend:

> People respond socially and naturally to media even though they believe it is not reasonable to do so, and even though they believe they don't think these responses characterize themselves. [. . .] Even the simplest of media are close enough to the real people, places, and things they depict to activate rich social and natural responses.

More recently, scholars interested in "human–machine communication" have made similar arguments about the anthropomorphization of communicative artificial intelligence (Guzman 2018; Guzman and Lewis 2020; Natale 2020; Strengers and Kennedy 2020). But unlike such devices as Apple's Siri or Amazon's Alexa, the personification of platforms like Netflix, Spotify, and TikTok relies primarily on algorithms and the design of interfaces rather than on communication cues such as the voice. In this sense, the personification of algorithms presents additional challenges to the media equation.

Personification is a cultural process through and through. In their experimental studies, Reeves and Nass (1996) found that people tended to be polite to computers and used contextual social rules to guide their interactions with them. Guzman (2018, 3) built on Carey's cultural approach to communication to argue that exchanges with technologies actually reveal "who we are, who we are to one another, and the very reality that we are creating." In a similar manner, as the following discussion of Spotify shows, users in Costa Rica drew on local and cultural ideas to personify the platform and its algorithms.

SPOTIFY, "MY LITTLE DUMMY"

As with Netflix, Spotify users both turned the platform into a reflection of their personality (by creating specific profiles) and also personalized it by treating it as an entity that had human-like characteristics (an interpellating subject). During focus groups and interviews, users tended to describe Spotify by using phrases such as "a little dummy" (*un muñequito*) or "toy" (*un juguete*), or "my little buddy." These expressions point to the importance of Spotify in providing both companionship and entertainment. Furthermore, when users wanted to emphasize its technological nature, they talked about Spotify as a "human brain." In this way, they associated Spotify with smartness and notions of artificial intelligence. This kind of definition also worked to impute particular motivations to Spotify. "Spotify functions the way we [the people] function," said Eliseo during a focus group, thus expressing his definition of Spotify as a reasoning person.

As figure 2.3 illustrates, the most common way to depict the platform during rich picture sessions was by characterizing it as having features such as eyes, hands, legs, hair, and even a (smiling) face. (See the appendix for an explanation of this method.) On some occasions, people made no distinction between how they represented humans and Spotify. When asked why she had drawn Spotify with hair, Sandra, a psychology graduate who created figure 2.3, responded laconically: "This is Sandra's Spotify (*el Spotify de Sandra*), thus it has my hair." With this response, Sandra meant that, on one hand, Spotify was her belonging (as in the Spotify *of* Sandra). On the other hand, she suggested that she was interacting with a particular version of Spotify (as in the Spotify *for* Sandra). By personifying Spotify, she thus fused various dimensions of personalization: she considered the platform both as an expression of her personality and a personal entity that constantly shifted to meet her specific requirements.

Her rich picture blended these ideas with eloquence. "Sandra's Spotify" is literally the center of a network of relations, media technologies, and activities. By drawing herself practically outside Spotify's operation and algorithmic calculations, she also expressed her belief that the Spotify subject operated intensively to offer her recommendations. It is for *her* (and nobody else) that Spotify works. In her drawing, she appears to respond to algorithmic interpellation in three emotional ways: by either liking, loving, or crying about a specific recommendation. Any one of these responses would cause her to play a song "on repeat."

How people thought of Spotify as a human-like being was perhaps nowhere clearer than in an exchange between participants during a focus group:

INTERVIEWER: How do you define Spotify? What do you think it is?

MARIANA: Someone very intense.

LAURA: Yes, a stalker.

GLORIANA: But I wouldn't want to humanize it. It is better to say that it is an online hacker.

NIDIA: Like a little ghost or weird little thing who says to you: "I saw that yesterday you were driving home and were listening to this. [Raises voice] I THINK YOU WILL LIKE THIS." And then just leaves. . . .

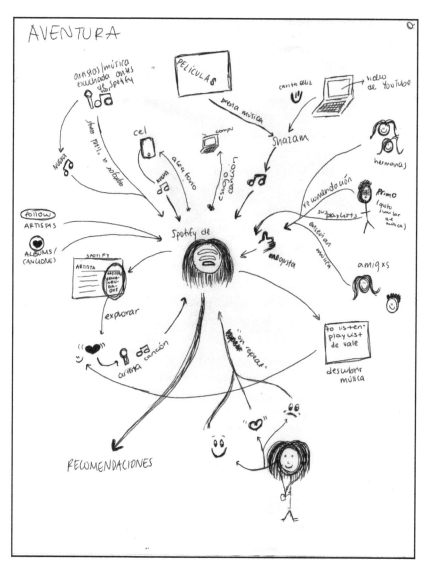

Figure 2.3
Sandra's rich picture of how Spotify recommends music. Drawings by "Sandra" courtesy of the artist.

When Gloriana realized that the group was personifying Spotify, she reacted by expressing a desire not to "humanize it." But the alternative she came up with was the notion of a person whose face is always hidden from plain sight.

In Spanish, the concepts of platform (*plataforma*) and app (*aplicación*) require using the article "*la*" and the female pronoun "*ella*." Thus, it could be expected that when users wanted to describe Spotify in those terms, they would use female pronouns. Yet interviewees strategically alternated between female and male pronouns to refer to different aspects of Spotify. They used "she" or "her" to describe Spotify's role in providing assistance to them. Users thus employed expressions such as "*she* suggests music to me," or "*she* organizes my music" (emphasis added). On other occasions, people employed male pronouns mostly to frame aspects of the platform they did not like. Accordingly, they criticized receiving unrequested algorithmic recommendations by defining Spotify as "a very annoying dude" or "the most intense of your friends (*amigos*)." Raquel, a 22-year-old business administration student, described Spotify as "a guy in his 30s who has become way too structured." By employing age stereotypes, she meant that, compared to previous versions of the app, Spotify now seemed too focused on recommending music and forgot about the features that made it interesting for her. Thus, Raquel used certain prevalent ideas and values in Costa Rican society to suggest she preferred Spotify as a "guy" in his 20s. Similarly, users invoked human-like characteristics associated with masculinity to question the motivations behind the platform's algorithmic recommendations. According to Nidia, "I know that *he* wants us to be friends because *he* is making money. *He* laughs with me, but *he* is not feeling anything behind that smile" (emphasis added). Nidia thus thought of Spotify as an insincere male counterpart who was primarily motivated by greed.

DEALING WITH A SURVEILLANT "BUDDY"

By personifying Spotify, users naturalized surveillance practices. Most people I interviewed employed metaphors related to the human senses to describe data extraction practices. Rather than talking about Spotify as a company, they represented Spotify in their rich pictures as an eye or used expressions such as "the eye that watches everything" and "the ears that hear everything." In these descriptions, they again employed male pronouns. Gloriana, a 20-year-old student, explained figure 2.4 in the following manner: "Spotify watches my life, *he* is there watching us. I've connected my [account] to Facebook, and [Spotify] thus has lots of information to process" (emphasis added). Gloriana's Spotify had legs, arms, and surveillant eyes magnified by binoculars. In this drawing, Spotify is able to watch everything Gloriana does: the people she talks to, the movies she watches, and the music she listens to on radio. It can even decipher the affective intention that motivated her to create certain playlists. The title of her drawing, "The Cycle of Recommendations," suggests this is a process she can't escape.

Yet, more than a "Big Brother," users like Gloriana thought of Spotify as a "Dear Brother." It was, after all, a "buddy." Spotify users thought that surveillance was necessary for improving their music recommendations. In this perspective, surveillance seemed justified. As Segura and Waisbord (2019, 417) have noted,

> in Latin America [. . .] the politics of data surveillance work differently than in the United States and other Western countries insofar as states historically did not develop massive, effective large-scale operations for gathering, analyzing, and managing data about populations during the past half century.

As noted in chapter 1, this observation is particularly salient for Costa Rica, which has built a national identity around the idea of peace since 1948—when the military was abolished (Sandoval 2002). In the absence of historical precedents to evaluate its consequences, surveillance seemed like less of a threat.

Moreover, users considered that surveillance endowed Spotify with great capacities. Like most users, both Sandra and Gloriana seemed satisfied with Spotify's surveillance posture in their rich pictures. As figures 2.3 and 2.4 show, they drew themselves smiling while enjoying "the cycle of recommendations." Pablo, an engineering student, explained:

> Spotify, the little dummy, asks: 'What would you like to listen to?' It knows my tastes a little bit and [adjusts] if I want something more intense, if I'm sad, or upset, or happy, if I want something more chill or something for any occasion. It *always* knows what I want. [Emphasis added]

For users like Gloriana and Pablo, Spotify acted like a psychologist who could identify their moods and desires and help them recognize their own emotions and affective states. Users responded to what they felt were appropriate algorithmic recommendations on Spotify the way they would react to a person:

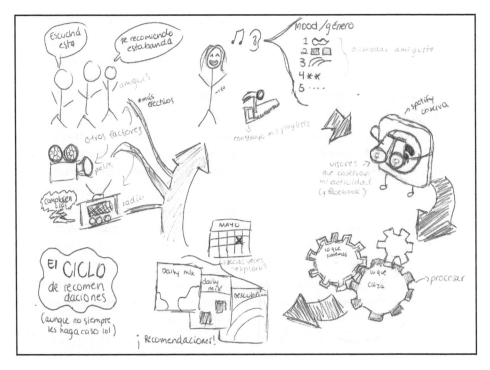

Figure 2.4
Gloriana's rich picture of how Spotify recommends music. Drawings by "Gloriana"
courtesy of the artist.

"I feel heard, valued," said Raquel. (As the next section on TikTok shows,
this feeling of appreciation was common in users of all platforms.)

SPOTIFY AS A SOCIAL INTERMEDIARY

Conceiving of Spotify as a human-like entity also involved attributing specific *social* capacities to it. Users thought that Spotify played a key part in
their most intimate social relations and daily activities by allowing users
to share music with others, learn what others were listening to, and talk
about music with them. Laura, an architecture student, explained: "[Spotify]
is very much ingrained in my social relationships. It influences a lot how I
interact with people around me." In this perspective, Spotify was a privileged
social intermediary: its algorithmic recommendations were constitutive
of social relationships. (This issue is addressed in more detail in chapter 5.)
Using a Facebook account to login into Spotify further solidified this view.

According to Raquel, logging through Facebook was like "giving Spotify permission" to be a part of her social life, to learn how she and her friends talked about music. As a result, she viewed Spotify as a musical mediator in her social network of relationships.

Users were convinced that Spotify took into consideration different data to make algorithmic recommendations, such as their music consumption practices (how often they listened to music and what kind of activities they performed while they listened to music); the moods and affective states surrounding these practices; and the particularities of the music itself (metadata about music genres, styles, tempo, etc.). People offered their own theory to explain how these factors were combined and turned into personal recommendations. In their logic, Spotify was able to recommend music because of its position as an intermediary of social relationships. Users thought that, in that position, Spotify could construct patterns of similarity with users who shared sociodemographic characteristics with them. They referred to this process as "averaging" data inputs. Leo, an audiovisual producer and self-taught musician, noted: "I imagine that what the platform wants is to *average* certain factors" (emphasis added). This interpretation suggested that, as a social intermediary, Spotify recommended the most typical or common music heard by particular social groups. In this way, the platform captured the musical preferences of most people. This is reminiscent of Cohn's (2019) view of algorithmic interpellation as a process of social comparison: it works by suggesting that certain recommendations are offered to people insofar as they represent or belong to certain social groups.

Together, the analysis of Netflix and Spotify presented thus far shows how personalization goes beyond a simple desire to get personal recommendations. Although this is certainly a part of the process, personalization also rests on the notion that users are being interpellated by a subject and on the attribution of human-like features to that subject. An additional part of this process is to establish a communication relationship with interpellating subjects. In what follows, I illustrate this dynamic with the case of TikTok.

BUILDING A PERSONAL RELATIONSHIP
WITH TIKTOK'S ALGORITHMS

Personalization culminates in the establishment of a particular relationship with algorithms. In this section, I examine how Costa Rican users understood the workings of TikTok's algorithmic content personalization, how they came to this understanding, and what the implications of their awareness are when developing a particular relationship with this app. I show that users' personal relationship with TikTok's algorithms was not constant but changed continuously: it shifted from experimenting, to training so-called "aggressive" algorithms, to "collaborating" with them, to enjoying the results of this collaboration, to "loving" them, and to taking care of them. Establishing this relationship builds on other personalization dynamics such as interpellation and personification.

Drawing on communication and media scholarship, researchers have recently begun to broaden the frameworks employed to make sense of the user–algorithm relationship. For example, Schwartz and Mahnke (2021, 1044) framed this relationship as a type of "communicative action." Considering the case of Facebook, they argued that

> users are communicating with Facebook's algorithm during everyday use rather than communicating through it as a simple medium channel, even though they might not think critically about everyday use as a communicative relation.

Cohn (2019) and Lomborg and Kapsch (2020) opted for Stuart Hall's classic notion of "decoding." They used this approach to explain how algorithms become meaningful for people. Cotter (2019, 900) referred to this as "playing the game," an interaction between people who are disciplined via algorithms but also play a role "in directing and making sense of their own behavior through interpretations of the game."

I propose to go a step further by examining how users' relationships with algorithms undergo a set of "passages" through which attachment is performed. In other words, it is not that users establish a relationship with algorithms that is either dominant, negotiated, *or* oppositional, to use Hall's terms; it is that they pass through moments of acceptance, negotiation, *and* opposition. Understanding people's relationship with algorithms thus requires a broader conception of agency that admits that people can simultaneously accept, negotiate, and resist recommendations. To make this case, I draw on actor-network theory and its approach to attachment.

ATTACHED TO TIKTOK

Attachment refers to how people and things become linked to each other. In Hennion's (2017b, 71) words, attachment is "what we hold to and what holds us;" it "signifies a connection, restriction, restraint and dependence" (Hennion 2017a, 113). Hennion emphasizes how agency is distributed in networks or assemblages of actors of different natures. From this perspective, algorithmic technologies such as TikTok are best understood as

> provisional results of a heterogeneous tissue of relations [i.e., assemblages] being continually tried out, tested, reshaped, in order to produce other objects, without being able to reliably distinguish content and support, network and actors, products and users. (Hennion 2016, 292)

The distribution of agency in assemblages is never final. Instead, it is constantly and dynamically performed. Through these processes, actors (both people and technologies) acquire specific capacities and roles. Gomart and Hennion (1999) refer to these agency shifts or redistributions as "passages." According to Gomart (2004, 87), rather than assuming "that there [is] from the start an already constituted subject," this approach focuses on the passages "through which 'subjects' [and objects] emerge, and emerge with new capacities."

Callon (2017) identified three mechanisms through which attachment is produced: dialogue (conversations devoted to showing that things provide a response to specific problems); coproduction (users' participation in the process of conceiving and developing an artifact); and addiction (a relationship of dependence that results in a quasi-mechanical attachment). In a similar manner, Verbeek (2005) argued that the attachment between people and technology develops through two main dynamics: "transparency" (the sense that users can understand and relate to an artifact's inner workings) and "engagement" (the notion that people can actually be a part of technologies' functioning).

I add maintenance to the list of attachment mechanisms outlined by Callon and Verbeek. An underlying premise in the study of maintenance is that fragility is a constitutive part of life and things (Denis and Pontille 2015). Maintenance is a process oriented toward the active prevention of decay. It challenges a view of technology as stable and envisions it instead as always "in the making." In this logic, permanent adjustments and transformations are technology's mode of existence. As Denis and Pontille (2020, paragraph 7, my translation) show, maintenance is a creative process: it "creates or strengthens bonds [. . .] [and] recognizes and concretizes both the interdependencies and the 'attachments' that hold together caregivers and objects/subjects of care."

In what follows, I examine the passages that led TikTok users in Costa Rica to the awareness of algorithms on TikTok. I show how this awareness created a context for understanding personalization in the TikTok assemblage, that is, how specific roles, relations, and capacities emerged and were constantly redistributed among users and so-called "aggressive" algorithms. I then analyze how the relationship between users and TikTok's algorithms underwent a series of passages through which attachment to the platform emerged.

PASSAGE FROM EXPERIMENTATION TO MUTUAL TEACHING

For most people in our focus groups, using TikTok began with a period of experimentation. Users reported having to figure out how the app worked and what it was all about. This process took time and practice. Experimenting with TikTok also required overcoming the "shame" (in their words) of using an app associated with teenagers and spending time in "nonproductive" activities. Some users said they abandoned or deleted the app after their initial frustration with it. The following experience described by one person is illustrative in this regard: "at first, it was hard for me to get a grip on the app, and I didn't like it. But when you personalize your 'For You' page, it becomes a catalog of everything you like. It's infinite and really never stops!" In this account, personalizing content became the goal of using the app and was the turning point in the user's relationship to it. Through experimentation, users acquired not only a better understanding of its workings but also a sense of how TikTok's algorithms operated.

Researchers have noted that awareness usually stems from active engagement with a platform (Eslami et al. 2015). This was also the case for TikTok users. They started noticing the operations of algorithms as they employed the app's features in certain ways. Nicolás (22 years old) explained a typical experience:

> [I learned] that, if you left a [video] "pressed," you could say to [the app] that you are not interested in it. That was great, because it was like, 'Hey, I can get rid of many things I'm not interested in!' It's one of those things that make you better understand how the platform works.

Active engagement thus helped establish the notion that specific user practices led to discernable results in TikTok's algorithms. In this way, users incorporated the premise that they were required to actively participate in making their interests and disinterests explicit. As for the cases of Netflix and Spotify discussed earlier in this chapter, users interpreted the behavior of TikTok's algorithms as a form of interpellation, as if the app were speaking directly to them and demanding that they be aware. Mónica's words were revealing in this regard: "[TikTok] *expects* you to teach it what you like" (emphasis added). Accordingly, users learned to employ those features that they considered most effective for producing a noticeable response from algorithms.

Users typically employed the notion of "training" to describe this relationship with algorithms. But the training went in both directions: users trained algorithms, while they also learned from algorithms how to do this training. Another common way to characterize this relationship was "depuration," the belief that users could clean out the impurities in their feed until it reflected only their interests. Rodrigo, a library science major, explained that "depurating" algorithms meant that he was taking "control" of the situation. Framing the relationship in this way made him feel that personalization was partly a product of his own decisions and actions. In this way, users claimed a sense of agency in how they related to algorithms.

My interviewees also perceived TikTok's interface as an affordance that guided their behavior with the app. Specifically, there was common agreement among users on the importance of the "For You" page as the starting place for interacting with content. For Nicolás, this was a product of how the app is designed: "[TikTok] leads you straight there. That's what you get first [as the default setting] and that's what you learn to use. Therefore, that's where I stayed." As Nicolás suggests, this affordance solidified the notion that personalized content is the default condition in TikTok. Users also made sense of the "For You" page by opposing it to the "Following" page (the space for watching content from selected users): whereas the latter allowed them to see what some specific contacts had posted, the former was interpreted as the domain of algorithms.

REACHING PERSONALIZATION

Through this passage from experimentation to mutual training, users gained specific understandings of the operation of TikTok's algorithms. They also developed a particular framework (or "context") to make sense of personalization on the app (Dourish and Bellotti 1992). Users believed that, to organize the abundance of videos, TikTok classified content by categorizing it according to formal criteria. They typically referred to these categories as "genres," "niches," or "segments." Personalization specifically meant the algorithmic assignment of appropriate "genres" to each person. Lucía, a 21-year-old law student, explained her theory in the following manner: "TikTok incorporates lots of genres; if you like one, the algorithm will continue to show it to you. It has informational pieces, tips, comedy, sad things, there is everything in there!" The clearest indication that users had normalized this view of personalization was a recurrent tendency to speak of "*my* algorithm" to account for this process.

Mónica used a metaphor that highlighted what she felt was TikTok's capacity to use personalization as a "trap" (cf Seaver 2019a): "I do think [TikTok's algorithm] is very aggressive in that it continues to show you what it already knows you like, it boxes you, boxes you, and boxes you by showing you only that!" In this account, personalization on TikTok constrained users (in both the sense of restricting and confining) within the boundaries of segmented content.

Given the centrality in Costa Rican society of the myths of being the most "peaceful" and "happiest" place on earth, the meaning of the term "aggressive" needs to be further unpacked. By "aggressive," users typically emphasized how determined (or assertive) they thought that algorithms were in achieving the goal of personalization. Although users also considered the capacity shown by algorithms to confront or interpellate them, they did not associate this aggressiveness with explicit forms of violence. Users also added a temporal dimension to their conception of algorithmic aggressiveness. In this view, being aggressive manifested in the speed with which algorithms could "learn" from users.

A view of TikTok's algorithm as extremely "precise" supplemented this definition of aggressive personalization. This view was expressed in the conviction that nothing about recommendations for users was coincidental. It was typical for users to combine ideas about personalization on TikTok and conceptions of its precision to define the app's most distinctive capacities. A telling example of this dynamic came when Mario, a 23-year-old international relations major, indicated that he had discovered a new type of "genre" that he really enjoyed without knowing it, and for which he had no name until TikTok revealed it. He referred to this newly discovered "genre" as "things with Vine-energy about mundane things that really make me laugh," thus comparing TikTok's content to the videos he often found on Twitter's service Vine (see chapter 3 for more on this comparison.)

Users also added a temporal dimension to their conception of algorithmic personalization. Valeria, who is 23 years old and currently unemployed, noted:

> TikTok's algorithm is brutal! You use TikTok three times, or maybe two, and it already knows everything. My 'Insta' doesn't know I'm queer [she used this term in English]. It never shows me anything queer. But by the second time I used TikTok, it was showing me lots and lots of queer stuff. That's the magic of the platform. I identify myself as *queer*, not *lesbiana*. My own mother is not able to understand this distinction. But TikTok stopped showing me lesbian things and began showing me things for queer people. I was like, "WOW!" [Emphasis added]

Valeria thus combined ideas of personalization (queer as a "genre" category), precision, and speed to describe how magical, surprising, and worthy of admiration TikTok's algorithms were.

Users not only gained awareness of the existence and capabilities of TikTok's algorithms but also developed a way of understanding their relationship with these algorithms. Put differently, users developed an awareness of the *joint* capacities of people and algorithms as an assemblage. Rodrigo offered a vivid illustration of this awareness:

> It's like working together. The algorithm has its formulas, its patterns, and its ways of reading my content, but [I can] pause a video and then play it again or go back and scroll down [in order to] influence the algorithm. I am aware that giving "likes" will influence my algorithm, and I use this awareness to give "likes" or not. In that sense, I'm collaborating.

Rodrigo's account favored the language of collaboration; he posited a view of users and algorithms working together toward the common goal of personalization. There was a clear distribution of responsibilities and capacities to achieve this end.

PASSAGE FROM MUTUAL TRAINING TO "ACTIVE PASSIVITY"

As users felt that TikTok's personalization had been achieved through "collaboration" (or from learning how to train algorithms), another passage typically occurred. Because of the "aggressive" nature of algorithms, users felt that they eventually could "leave everything in their hands," as Catalina (a public relations specialist) described it. During those moments, training was no longer necessary: users allowed themselves "the right to be happy" (using Catalina's own words), to intensively experience positive emotions on TikTok without having to interfere. Catalina continued: "If I see something that shouldn't be there [. . .] I don't even bother. My mind thinks, 'The algorithm will take care of making it disappear.' And that's it." Mario described arriving at a point where failures in algorithmic recommendations were so rare that no major intervention was required.

The passage to "enjoying" personalization on TikTok should not be confused with inaction. It is more about "active passivity" in Hennion's sense: "it is not about moving from activity to passivity, but to act in order to be acted on. Things have to be done, to make things happen [*pour que les choses arrivent*]" (Hennion 2013, paragraph 29, my translation). In this sense, this passage is built on a sense of achievement: it was a way of being entertained by the results of past "training" and "depuration" efforts. It was the time to reap the "harvest," as Rodrigo defined it.

In this passage, the user–algorithm relationship was enacted as an emotional matter. Users shifted their vocabulary from notions of "control," "training," and even "collaboration" to that of mutual "love." Catalina explicitly stated her feelings for TikTok's algorithms: "I shaped it and now I love what I have, so I definitely think it's a super aggressive algorithm, and I love that it is aggressive!" As Hennion (2017b) reminds us, attachment is reciprocal. Thus, the "love" for algorithms also had a counterpart; it evoked a feeling of appreciation that Laura captured with precision: "Since I hardly see this [personalization] on other social media [platforms], and I know that TikTok gives me content just for me, I feel special, I feel that it's giving me attention, making things just for me."

PASSAGE FROM PLEASURE TO CARE

The TikTok assemblage also needed to be maintained to keep the "mutual love" going. A new passage was required to preserve TikTok's algorithmic personalization from decay. I refer to this specific passage as "care." Borrowing from Denis and Pontille (2015, 360), a focus on care invites a consideration of TikTok as a "fragile and mutating entit[y], [. . .] [a] thing that ha[s] to be taken care of, despite [its] standardized design and despite [its] ordering aims." Care is also a means to channel user resistance to unwanted content (such as viral trends that are explicitly sexist). Tania, a college student, thus employed the vocabulary of love to suggest that, when this happened, her relationship with TikTok could feel "toxic."

Care began by mobilizing awareness in strategic ways. For Rodrigo, there were times when he felt that "his algorithm" required some work to "re-reroute" his content. These words illustrate how people sought to incorporate awareness as a regular component of usage practices. This awareness then translated into specific roles. It was common for users to employ expressions such as being "careful" (*cuidadoso*) or "cautious" (*cauteloso*) to define the role they needed to perform in this process. Valentina's words are worth quoting at some length:

> The algorithm is definitely something I think about whenever I interact with the app. This is the app where I am most careful about what I give "likes" to and what I "hide" [. . .] because I don't want anything to be added to my algorithm that doesn't interest me, because it [the algorithm] seems to me to be too effective. Thus, a misplaced "like" can mean that my algorithm could get disarranged [*se desacomoda*].

Like Valentina, users incorporated their awareness of TikTok's "aggressive" algorithms (in the sense of being extremely precise and responsive) to maintain balance in the assemblage. Being "careful" meant conceiving the "aggressive" nature of TikTok's algorithms as something that needed to be dealt with continuously to avoid the decay of personalization. The logic of care was thus one of preservation (hence the idea of "disarrangement") rather than "depuration," "training," or "collaboration."

This carefulness manifested in a series of specific practices and ways of relating to TikTok's algorithms. All the knowledge gained during the "training" and "depuration" of content had to be practiced with discipline. Daniela described this discipline as "conscious effort." She explained that she needed to keep performing certain actions so that TikTok could "understand" that she didn't like certain types of content, such as "TikTok dances." These words suggest that users continued to rely on the notion of TikTok's "aggressive" personalization through "genres" to indicate how they thought the app could be maintained.

PASSAGE FROM CARE TO "ADDICTION"

A final dynamic in the process of attachment was the recognition of "addiction." Self-admissions of addiction to TikTok were common during the focus group discussions. When they employed it, users did not necessarily mean a literal physiological need to use TikTok. Many claimed that they actually didn't use the app every day or used it only when they knew they had time to spare. Yet they agreed that their addiction to TikTok was distinct from their attachment to other platforms. Users typically used "addiction" to express two ideas: how they tended to lose the perception of time when using the app, and how difficult it was to disrupt its use because of the pleasure it produced. In these accounts, addiction meant having to self-regulate the use of the app to avoid spending more time than what seemed reasonable.

Users mentioned several reasons to account for their addiction to TikTok: the memetic value of the content, the short format of videos, the overall app design, the infinite scrolling, the ease and allure of the swipe gesture, and the availability of time to use the app while they stayed at home during the COVID-19 pandemic. In short, they believed they were addicted because personalization worked too well. To be sure, such statements confirm the addictive nature of contemporary technological design (and the dopamine rewards it produces). But following Lovink (2019) and Gomart and Hennion (1999), I treat them also as expressions of the particular kind of attachment that users had established with TikTok through the process I have discussed in this chapter. Seen in this way, addiction also means force of attachment. Put differently, addiction to TikTok was achieved through the repetition of passages between experimentation, mutual training, active passivity, and careful practice. Georgina, a 22-year-old psychology student, expressed this idea eloquently:

> I do think that, to a certain extent, [algorithmic] aggressiveness is what makes you so hooked to TikTok. I know of no other app that [recommends] things to you in such a direct way, and that is so easy to use. You just 'swipe' like this [mimics the gesture] and you get all the content you want.

Georgina's account provides an ideal summary to the process I have discussed thus far. She began by implying the importance of awareness for understanding how a relationship with algorithms forms. She then acknowledged the relative "aggressiveness" of TikTok's algorithms in responding to conscious user practices. And she concluded by alluding to the swipe gesture as the embodiment of her attachment to TikTok (rather than its cause) and as a symbolic preparation for the personalized feelings and emotions that certainly come next. The continuous process of consciously performing and maintaining this assemblage, over and over, time after time, is the addiction.

CONCLUDING REMARKS

Personalization dynamics illustrate what Vaidhyanathan (2011, 110) referred to as "infrastructural imperialism": that is, how platforms try to "structure ways of seeking, finding, exploring, buying, and presenting [data] that influence (though they do not control) habits of thought and action." This form of influence, Vaidhyanathan argued, is much more profound than traditional cultural imperialism, in that it helps normalize settings (such as the personal profile) as the default way of interpreting culture and the self. In this way, motivated by the promise of getting personally relevant recommendations, users incorporate the precise mechanism through which technology companies also seek to domesticate them: the segregation of consumption practices into compartments that companies use to influence users' behavior.

To supplement such dominant accounts in the scholarly literature that emphasize the operation of platforms and algorithms in shaping user practices, this chapter has put forth a view of personalization as a communication process. I have argued that personalization is fundamentally about establishing a *personal* relationship between users and platforms through a series of dynamics. This chapter has discussed three of these dynamics: interpellation, personification, and passages in the user–algorithm relationship.

Through interpellation, users experience algorithmic recommendations as efforts to address them in a personal manner, as if someone were calling their names. This experience is reinforced by the design of interfaces that work to convey the notion that all communication is addressed to *you* and to no one else. Considering the case of gendered recommendations suggests that markers of identity are key for the interpellation process (and personalization more broadly). Personifying platforms runs parallel to interpellation. As users believe that they are being addressed in a personal manner, they try to make sense of the subject that interpellates them. Drawing on cultural rules of social interaction and gendered stereotypes, they thus attribute some person-like features to platforms. Finally, the process of personalization culminates in efforts to establish a unique relationship with an interpellating subject. Users perform different kinds of roles as they interpret continuous changes in algorithms. An attachment to the platform is established as the relationship between users and algorithms undergoes a series of passages.

The tension between the local and the global informs the dynamics of personalization discussed in this chapter. On one hand, the users that I studied often forced algorithms to comply with traditional Costa Rican values and ideals in order to judge the algorithms' personal relevance or make sense of their significance. Thus, users turned to local rules associated with gender performance or to age stereotypes that are prevalent in Costa Rican society as they personified platforms such as Netflix, Spotify, and TikTok. On the other hand, algorithmic interpellation works best when it allows people to express (to others and themselves) that they could relate to cultural phenomena that they associated with the global north, primarily the United States (such as being able to value Hollywood "classic" romantic comedies from the 1990s or understand the meaning of queerness, using the English term to name it.) Chapter 3 elaborates on this cultural tension in more detail through the notion of integration.

The present chapter also has revealed the centrality of genres in personalization. On Netflix, both traditional and alternative genres are key in sustaining interpellation dynamics. Bundles play a central role in linking algorithms to the interpretive contracts materialized in genres. Likewise, TikTok users consider that the algorithmic recommendation of specific genres is the clearest sign that successful personalization had been reached. Although the analytical focus has been on the relationship that people establish with socio-technical systems that draw attention to certain types of content, the chapter also has shown how technology and content are interwoven in processes of personalization. Put differently, personalization is a "texto-material" process (Siles and Boczkowski 2012).

Although discussed separately, interpellation, personification, and passages work simultaneously. Moreover, these three dynamics characterize how users enacted algorithms on Netflix, Spotify, and TikTok. Because of the perception that TikTok's algorithms were more "aggressive," personalization was arguably more intense on this platform for the users that I studied. These users argued that personalization occurs faster and in a more precise manner on TikTok compared to Netflix and Spotify. This difference of degree can thus be explained as an outcome of mutual domestication: it involves both algorithmic operations and specific user practices to bring about personalization.

This approach to personalization helps broaden the understanding of why people interact with algorithms in certain ways and provide their data to technology companies. Users certainly personalize their profile to receive relevant algorithmic recommendations. But they also do so to sustain a communication relationship with a person-like being that *demands* their attention and reaction, to which they are also attached and that attaches them to other people (see also chapter 5). Creating and maintaining a profile is a way to respond to subjects that interpellate them by asking them: "who is watching?" Put differently, the profile is users' response to this question: this is "me." This person at this place and time. To make sense of the subject behind interpellation, users typically draw on cultural references and parameters to guide their interpretations. For example, users assess technology companies' surveillance practices in their own sociocultural history (in this case, Costa Rica).

As noted at the start of this chapter, at stake in the process of personalization is the agency of users. By integrating evidence from these three cases, I have argued for considering how agency shifts constantly through an evolving relationship with algorithms. In this perspective, those occasions when users simply follow algorithms are best understood as instances of "active passivity." Users have participated actively in reaching such a state of passivity and will continue to work intensively to prevent the decay of personalization. Thus, for many users, personalization is a cultural achievement that requires constantly and variously weaving a relationship with algorithms. Chapter 3 examines the variety of cultural resources and repertoires involved in developing this relationship.

INTEGRATION

Choice is at the core of users' relationship with media and technology (Has brink and Domeyer 2012; Salecl 2011; Webster 2014; Webster and Waksh-lag 1983). Choice has also been a concern for scholars who focus specifically on datafication issues (Danaher 2019; Gal 2018). Authors have argued that algorithms matter because they perpetuate the neoliberal tendency to make individuals perceive themselves as choosing subjects. For example, drawing on Althusser (2014), Cohn (2019) suggests that algorithms help reproduce forms of capitalist dominance by continuously legitimating the focus on individual choice. In a similar manner, Kotliar (2021, 364) noted that algo-rithms are "incessant generators of choice that not only aim to produce different kinds of choosers but that also recursively corroborate choice's role in contemporary life." Research has emphasized how algorithms guide users toward preestablished options that serve to reinforce their sense of "individuality, self-worth, and authenticity" while operating instead as mechanisms of control (Cohn 2019, 7). For this reason, authors have con-cluded, algorithms only intensify the illusion of free choice (Sadin 2015) or the "myth of choice, participation and autonomy" (Arnold 2016, 49).

This chapter nuances prevalent accounts of algorithms' influence on user choices. Concurring with Cohn (2019, 8), I argue that "in practice, the controlling, interpellative effects of recommendation systems are far from total." In short, I show that algorithms do not work entirely alone, nor do they act as the only determinant of choice. Instead, this chapter shows how users integrate algorithmic recommendations into a matrix of sources, capacities, and relations based on their sociocultural backgrounds, the situ-ations they want to resolve, and the material conditions of their digital environments. The notion of integration emphasizes the work carried out by individuals to combine a variety of cultural elements as they ponder what recommendations they want to follow.

This view of cultural integration can be compared to Grounded The-ory's approach to theory building or the linking of categories that leads to conceptual constructs that fit with the data (Corbin and Strauss 2015). In Grounded Theory, integration is a work of "reconciliation" through which researchers consider multiple and overlapping realities expressed by the data (Corbin and Strauss 2015, 296). In a similar manner, users integrate overlapping elements and realities in ways that fit with their sociocultural contexts, situations, and expectations. Borrowing from Morley's (1986, 14) classic work on audience interpretation, the integration of algorithmic recommendations needs to be situated within "the socio-economic structure of society, showing how members of different groups [. . .], sharing different 'cultural codes,' will interpret [media] differently." The analysis presented in this chapter works to further situate life with algorithms within the structure of Costa Rican society.

The chapter discusses three types of integration evidenced in the use of different platforms. I begin by examining how Costa Rican users combine various types of cultural *sources* or criteria as they choose what content they want to watch on Netflix. Considering the case of Spotify, I then discuss how users in the country choose algorithmic recommendations as they seek to integrate certain *capacities* in their lives: to be a specific kind of person, to negotiate a sense of belonging in certain social groups, to sustain or strengthen ongoing social relations, and to enact the local/global tension. Finally, I analyze how TikTok users integrate their experiences with various media technologies to make sense of algorithmic recommendations on this app. They specifically appropriated TikTok as a platform that differed from others but at the same time was similar to, retrieved, or even influenced other technologies.

INTEGRATING CULTURAL SOURCES TO WATCH NETFLIX

How and why people select content has been a key concern in the social sciences and humanities for several decades (Heeter 1985; Webster and Wakshlag 1983). Webster (2014, 27) summarizes some of the main conclusions derived from this body of work by suggesting that "as a rule, researchers have assumed that users will select what they like and avoid what they don't like." In practice, Webster (2014) argues, this assumption has translated into a growing number of studies on people's construction of repertoires (the limited number of outlets that people rely on for content) and heuristics (the rules of thumb they apply to decide, such as fulfilling expectations and obtaining social approval).

The process of deciding what content to select has become more complicated, due in large part to the abundance of sources that users have access to (Boczkowski 2021). In this first section of the chapter, I look at how users in Costa Rica navigated this context of apparent endless choice by situating their relationship with Netflix and its recommendations in a wider repertoire of cultural sources. I describe five types of cultural criteria that people have integrated into their lives: interpersonal relationships, technical characteristics of the content, the role of Netflix as a producer of original content, reviews in specialized outlets, and opinions available in social media. I then discuss how users combined these sources into repertoires that helped them respond to specific situations and demands in their daily lives.

The perspective of opinion leaders, that is, the thoughts of people whom users value and respect, proved to be a decisive criterion for some users. For example, Catalina, an audiovisual producer who runs her own company, pointed out that the number of times she would watch something that Netflix recommended to her was "tiny" compared to the number of times she did it because someone else suggested it. Users accounted for this preference in two important ways. On one hand, they argued that Netflix could not know them as well as the people in their social networks. They thus turned to specific people for recommendations. Inés, a 25-year-old miscellaneous worker, explained it like this:

> I have two or three friends who are also super fans of movies and series. So, I pay a lot of attention to them because they have already recommended things to me and they've been very accurate. If they have good things to say about something, I know I'll like it. They can also say: "It's bad, but you still have to watch it," and I'd still watch it.

For her, recommendations from a select group of friends seemed more trustworthy than from algorithms because they were grounded on a history of shared experiences. They have also proved successful in the past. Therefore, she was even willing to watch content that was labeled as "bad," as long as these friends guaranteed that she would enjoy it.

In addition to shared history, users also trusted other people over algorithmic recommendations because they came from an embodied experience. That was the case with Ema, a 39-year-old college instructor, who noted: "Other friends, or colleagues, even my own students have sometimes recommended that I watch a series. Because they do watch many!" Ema trusted both strong and weak ties in her social network of acquaintances because they spent more time than her watching content. Their experience thus operated as a filter that she could rely on when considering options.

On the other hand, users also argued that they preferred the recommendations of people over algorithms because they couldn't know Netflix as well as they knew their opinion leaders. These users perceived Netflix's recommendations as commercially motivated. Consistent with the notion of algorithmic as a "very annoying dude," discussed in chapter 2, many users expressed the opinion that Netflix was "too intense" in recommending its own series. These words build on personification dynamics to suggest that Netflix's recommendations should be viewed with some suspicion.

Formal specificities, such as production characteristics (for example, who the leading actors are, who the production team includes, or the presence of a particular director or cinematographer), are additional factors that shaped the consumption of the platform. By providing abundant information about the production characteristics of the content available on the platform, Netflix makes it easy to implement this criterion. Referring to one movie that was displayed on her profile during the interview, Ema noted: "We watch many movies with Adam Sandler. I haven't seen this one but [would watch it because] Adam Sandler is in it." Ema thus created a content category ("Adam Sandler movies") she would watch that is based entirely on the presence of an actor. Because the algorithmic bundle had aptly captured this interest, she said she would be more willing to watch this recommendation.

Traditional genres are another instance of this criterion. As noted in chapter 2, genres remain an important source for people to assess and manage expectations about content. As a criterion, users connected expectations about genres to the idea of "taste." In other words, they associated the practice of watching a specific type of content with the notion of being a certain kind of person (defined by their taste). Erick, a 38-year-old photographer who shared an account with his girlfriend and daughter, tied his consumption of certain genres to his adherence to traditional gender roles. He noted: "The genres that I like are action, comedy, horror, and science fiction. Everything that is not drama or romantic, cheesy movies. That's not for me. That's something you'd see in my girlfriend's [Netflix] list." For Erick, following a recommendation meant protecting his masculinity (in the sense that he interpreted dramas and romantic movies as appropriate suggestions only for his girlfriend.) Users thus privileged recommendations that allowed them to bolster the sense that they were cultivating their taste and their own selves in particular ways.

Some users also valued Netflix's role in creating and developing original content. Netflix displays this content in a specific section of its interface and includes the company's logo and the phrase "Netflix Originals." This section also shows large, personalized images, each show's logo, and a notification when new episodes are available. In this way, it becomes easier for users to identify and consume these recommendations.

Some users interpreted the production of original content as a sign of the company's evolution. Giancarlo, a political advisor, compared it to what he thought was the symbol of quality television, namely, HBO:

> Netflix made an amazing concept change and stopped being a streaming platform to become a television channel without being on television. They copied HBO's model without being on cable [television]. I think that's excellent because, in addition to a video depository, now Netflix makes its own content and I don't feel the price has changed. It is as if you were paying for a Premium subscription.

In this way, algorithms are integrated into an imaginary that emphasizes the value of Netflix's brand to produce and distribute its own content. As Bucher (2018) notes, imaginaries blend both ways of thinking and feeling about algorithms and shape users' attitudes toward platforms. Giancarlo welcomed this evolution of the company's identity, especially as it didn't translate into a price increase. Thus, "Netflix Originals" recommendations felt to him like a bonus, almost a gift.

Many interviewees associated Netflix with the notion of quality and thus positively received most recommendations about its original productions. Antonio, a 23-year-old clerk at a public university, maintained that, while he did not necessarily like all the plots or genres available in Netflix's "Originals," he thought they were all very well made. For Antonio, Netflix's own productions were characterized by a common aesthetic standard that guaranteed the quality of the content. Moreover, they stressed that Netflix made sure to satisfy a variety of desires in its audience by providing abundant and diverse options.

As noted above, the abundance of products to watch is one factor that complicates the selection of content for several people. For many users, the expertise of certain sources made a difference when choosing what to consume on Netflix. This criterion is different from opinion leaders in that it doesn't privilege the embodied experience of a trusted friend but rather emphasizes the source's formal credentials to prove the validity of recommendations.

Many interviewees said they turned to reviews of films or series on specialized websites to learn about recommendations. That was the case of Augustina, who is 58 years old and works as a university administrator. In her words:

> I like good movies, so I look for some that have been well received by critics. I like to do my research. Some movies have gained recognition from [awards], not only the Oscars but also Sundance, or because they were in different film festivals. I look for those types of films.

Agustina proactively looked for information to make an informed decision about what to watch on the platform. Cultural proximity with the United States was important to her when making a choice. It was the opinions of experts from American institutions (such as websites, awards, and festivals) that allowed her to define what a "good" movie was. To be sure, this does not prevent Netflix from reaching users with specific recommendations, as the company has repeatedly indicated that it also considers critics' reviews to improve its algorithmic recommendation features (Amatriain and Basilico 2012b).

Finally, an overwhelming majority of the interviewees considered opinions expressed in social media as a source of indirect recommendations. The experience of Manuel, a 36-year-old computer scientist, helps illustrate how users turn to social media to this end. During an interview, he explained:

> To be honest, I do care about what others are interested in and what they are watching. If a friend of [my wife], for example, publishes on Facebook that she is watching a program and why she likes it, and if I receive this alert and it makes me curious, I will watch it immediately and then try not to see more comments about it to avoid spoilers.

Social media thus provided users with recommendations through unsolicited comments about specific content that was gaining popularity. Users like Manuel felt they needed to watch content that became popular in their social networks because, otherwise, they might be isolated. These comments on social media then created pressure to watch suggested content quickly, because there was always the possibility that others would spoil plot elements before the person had a chance to watch the show.

On other occasions, users explicitly employed social media to ask for recommendations. Sandra, a 24-year-old university student, noted: "Sometimes I write on Facebook or Twitter: 'Hey, what series do you recommend on Netflix?'" Answers function as votes and as a way to assess criteria about specific recommendations. Whether this content was originally suggested to their contacts by algorithms is usually unknown to users but, by asking them, they feel these recommendations come from a trusted source (an embodied experience, as noted previously).

BUILDING SITUATED REPERTOIRES OF RESOURCES

Users integrated these recommendation sources in different ways and with varied importance. In this way, they created their own repertoires and hierarchies of criteria. The notion of repertoire has been used to characterize the "small subsets of the available options that people use time and again [which] simplify search and decision making" (Webster 2014, 37). In short, it has been employed to designate ensembles of media outlets (Haddon 2017; Hasebrink and Domeyer 2012; Taneja et al. 2012). Here, I use it instead to refer to the subsets of cultural resources and criteria that people draw on and combine to choose content on multiple platforms, including Netflix.

Interviewees drew on these cultural repertoires as a product of their sociocultural condition. They specifically integrated cultural resources in their decision making to respond to the demands associated with their work, personal relationships, and affective states. Users assessed the relevance of algorithmic recommendations according to these cultural repertoires based on how much these recommendations provided a fitting response to such kinds of demands.

The most obvious example of how occupations created opportunities for integrating cultural resources came from interviewees who worked in media-related industries. Ariana, a 37-year-old woman who works as a producer in the audiovisual industry, noted how social conversations quickly turned into professional exigences for her:

> The people that I work with are constantly talking about things that are coming out, so it is a social conversation that turns into work. I don't just watch what catches my attention, but I see everything that is coming out. Am I able to do it? No. But I do force myself to try to watch what comes out because it is part of my job.

Ariana thus integrated both conversations with peers and algorithmic rec-ommendations, because she felt this would allow her to fulfill her pro-fessional obligations in a better way. However, this was also the case for people who worked in jobs unrelated to the media. Sofía, a 41-year-old university faculty member in gender studies, thus said that she was willing to watch romantic movies recommended to her on Netflix because they could provide useful examples in her class to discuss with students. Thus, participation in certain social and professional spaces created conditions for accepting certain algorithmic recommendations.

Users also integrated sources from their cultural repertoires when they sought to establish a symbolic connection with someone else. Most sources seemed like valid options to meet this goal. Ariana, the movie producer, thus noted about a particular Netflix original series that was recommended on her profile when I interviewed her: "I would watch it just because my mom wants to watch it." Similarly, María, a 20-year-old university stu-dent who said she mostly valued recommendations from trusted friends, explained why she would make an exception in her practices to consider an algorithmic recommendation that was displayed on her profile during our conversation. When she saw the category "British Movies," she paused to explain: "My father is a fan of British cars, and so is my grandfather. Not only cars, but the culture as well, the machines. That is truly important in my life. I was shaped by this." A symbolic connection with members of her family thus enabled her to integrate algorithmic recommendations into her criteria, despite her initial resistance.

Drawing on research on the psychology of media choice, scholars have also emphasized the importance of certain predispositions in shaping peo-ple's selections of media and their content (Knobloch and Zillmann 2002; Zillmann 2000). In the case of Netflix, some users assessed resources in their cultural repertoires that could allow them to manage affective states (such as moods and emotions) through the selection of particular content. Ema, the college instructor, explained that she made some decisions regard-ing what she wanted to watch on Netflix depending on her mood rather than on algorithmic recommendations. In her words: "It's more about my mood than about saying, 'Netflix says that I should watch this'."

The case of Netflix suggests that algorithms do not work entirely alone. Instead, I have argued that users integrate them as part of repertoires of sources and criteria to deal with various situations and demands, depending on a matrix of social, cultural, and psychological codes. Some of these resources are not entirely new. Researchers interested in media users over the past decades have also stressed several similar factors that account for why people choose specific media outlets and contents. As Webster (2014, 14) notes, "the structural features of social life [. . .] aren't swept away with each new technology," including algorithms. As this study of Netflix use reveals, an analysis of how algorithms shape the choices of users also needs to account for how recommendations interact with the structure of everyday life in Costa Rican society.

INTEGRATING CULTURED CAPACITIES THROUGH SPOTIFY'S RECOMMENDATIONS

A second integration dynamic consisted of acquiring specific cultured capacities through the relationship with algorithms. Users followed algorithmic recommendations insofar as these allowed them to obtain capacities that they valued for resolving specific situations. Spotify offers an ideal case to examine the integration of capacities into people's lives because it ties music consumption to algorithmic choice.

To account for this process of acquiring cultured capacities, I draw on the work of Ann Swidler (1986; 2001b). Swidler (2001b, 24) departs from the notion of culture as a unified monolith shaping people's actions and defines it instead as "a bag of tricks or an oddly assorted tool kit [. . .] containing implements of varying shapes that fit the hand more or less well, are not always easy to use, and only sometimes do the job." In Swidler's approach, a cultured capacity is a set of practices and skills that people use to act in specific circumstances or contexts. Swidler enumerated the types of capacities that culture provides: to perform specific identities, internalize habits, negotiate belonging to social groups, and express certain worldviews. In Swidler's (2001a, 71–72) words, "Culture equips persons for action both by shaping their internal capacities and by helping them bring those capacities to bear in particular situations." Swidler's method is to look at the cultural resources available to people and the various ways in which individuals use culture to resolve situations.

In short, cultured capacities provide people with certain resources to act. Swidler captured the dynamic of acquiring these capacities through the notion of "strategies of action." She defined these as "general solutions to the problem of how to organize action over time, rather than specific ways of attaining particular ends [. . . .] [They] provide [. . .] one or more general ways of solving [. . .] difficulties" (Swidler 2001a, 82–83). Strategies of action are ways of resolving problems that derive from cultural experience.

Building on Swidler's approach, I argue that the Spotify users I studied chose certain algorithmic recommendations not only because of the musical content that was offered to them but also because they sought to develop cultured capacities that allowed them to resolve certain situations. In what follows, I show how users in Costa Rica turned to Spotify's recommendations to acquire or strengthen four specific capacities: to be a specific kind of person, to negotiate a sense of belonging in certain social groups, to strengthen ongoing social relationships, and to enact a position in the local/global tension. Some of these capacities are similar to those previously identified by Swidler, but others are specific to the case of algorithmic platforms' users in Costa Rica. (As with other chapters, although I examine these dynamics by drawing on examples from one specific app, they also apply to the other platforms I studied.)

PERFORMING IDENTITY

One of the most prevalent reasons that determined whether a user followed an algorithmic recommendation was the issue of identity. Users thought of Spotify and its recommendations as a way of performing a self, of being or becoming a certain kind of person (cf DeNora 2000; Hesmondhalgh 2013). During an interview, Rubén, a 39-year-old social psychologist, described his relationship with the platform in the following way: "[Using it] suggests that you can have a refined taste, it is not only for others but also for yourself." Rubén thus considered that specific parts of the self (such as "taste") were at stake in how he used the platform. Behind this assertion lies the premise that Spotify is embedded in social relationships and, as a result, it provides a window for others into the most private aspects of the self. (See chapter 2 for an explanation of how Spotify users envision Spotify as a social intermediary of interpersonal relationships.)

Many Spotify users in Costa Rica integrated recommendations into definitions of the self. Rubén's case provides once again a vivid illustration. Whether to follow an algorithmic recommendation was tied to how much it allowed him to perform what he thought was his most defining identity trait: to always distinguish himself from others. He exemplified this trait by referring to how he behaved in his two workplaces (a private university and a public institution): "When I work in the private sector, I am a socialist. But when I work in the public sector, I am a liberal. I have to take the opposite stand. I do not like being a part of the established order." Rubén then integrated this self-understanding into his relationship with Spotify. Accordingly, he explained that his main criterion for deciding whether to follow a recommendation was assessing the extent to which it would allow him to maintain his desire to "be different" from others.

It was also common for interviewees and participants in focus groups to account for their ties to Spotify by invoking their academic major, profession, or trade. Rubén thus argued that his relationship with Spotify expressed "a very social psychology thing." Many interviewees went even further by defining the platform in terms of meanings acquired in their academic training. For example, Rubén defined Spotify as a "feedback control system," a term he indicated came from the field of electrical engineering. Gabriel also noted that his notion of Spotify as a "system" was a product of his training in political science. In this way, definitions of the platform became an expression of who the person was and how they wanted to be seen. To be a "good" psychologist or engineer was to think of the platform in certain ways.

However, it would be misleading to conclude that people relate to Spotify and its algorithms based exclusively on their profession. Instead, I argue that ideas associated with specific majors (such as skills and professional routines) allowed users to bring specific capacities to bear in certain situations, such as defining themselves by how they appropriated the platform. Established understandings of professions provided them with useful symbolic resources to obtain self-forming capacities.

NEGOTIATING GROUP MEMBERSHIP

Algorithms also operated as a way to negotiate group membership. This is by no means a minor issue in Costa Rica, where more importance is usually placed on "group affiliation (as opposed to personal achievement)" and where "interpersonal bonds are highly valued" (Rodríguez-Arauz et al. 2013, 49). The premise in the acquisition of this cultured capacity was that knowledge of certain phenomena (bands, styles, artists, etc.) was shared by all members of a group. Adopting those phenomena thus becomes a way to signal membership.

Developing this capacity was of key concern for many Spotify users I interviewed. For example, for Nina, a 52-year-old audit specialist, believing that Spotify's recommendations averaged the collective preferences of others was crucial to her decision on whether to accept an algorithmic recommendation. She explained: "[I began using Spotify] maybe to not feel so outdated that sometimes I have to ask what young people are doing these days. They [young people] give me a lot in that sense so I don't stay behind." In this way, she argued, the platform allowed her to "understand what today's tendencies are" in music consumption. This explanation combined both matters of content and issues of technology, that is, an idea of how algorithms capture certain content (the music of "young people") and a belief of how this was achieved (averaging what most users listen to into group "tendencies") (cf Siles and Boczkowski 2012).

To be sure, people have traditionally seen music itself as a means of being part of certain groups (Hesmondhalgh 2013). But how users achieve this capacity of group membership now rests on how they specifically think that Spotify's algorithms work. For example, some users said they had learned about how algorithms operate specifically to improve their chances of receiving content they could in turn recommend to other people. Others took a very proactive role in using features that allowed them to explore the bands, artists, and styles they associated with certain groups. To that end, they typically employed features such as "Similar artists" or "Fans also like." One interviewee explained: "I used the 'Artists like Joy Division' and got to John Maus. Then I looked for 'Artists similar to John Maus.' From there you build the sequence, iterating and iterating, and you get to know lots of [artists]." In this way, users said they could be "ahead of the game" and learn references they could then recommend to others in their social groups to acquire or establish their position in the group.

STRENGTHENING SOCIAL RELATIONSHIPS

Related to the above discussion, users also followed (or not) algorithmic recommendations to foster the capacity of keeping or strengthening social relations that were meaningful to them. This capacity is aptly captured by Levy's (2013, 75) notion of "relational big data": "people *constitute* and *enact* their relations with one another through the use and exchange of data" (emphasis in original). Considering social relations as central to music consumption was key in relating to Spotify and its algorithms. Nina, the audit specialist, explained:

> I prefer the personal over the digital. I feel like I can put a face [to recommendations]. That's a trigger for me. It makes me say, "I have to take a look at this song because it was someone [who recommended it]." Emotionally, it's not the same.

Conceiving of Spotify as an intermediary of social relationships (a practice established through the personalization of the platform, as explained in chapter 2) thus created fertile grounds for users to accept its recommendations. In this way, users felt that Spotify offered them material for conversations that could help them connect with others.

A specific instance of this dynamic was using Spotify to maintain certain social relations through music. Many users indicated that they began listening to certain artists or songs because this music reminded them of those who helped discover it. Thus they acquired the habit of listening to certain artists because they were exposed to them through another person. In these cases, the music stands in for the relationship as a form of "inheritance." To develop this capacity, users assessed algorithmic recommendations based on the extent to which these could help them keep musical inheritances and relationships alive.

ENACTING LOCAL AND GLOBAL RELATIONSHIPS

Finally, the decision of whether to follow an algorithmic recommendation or not was also tied to the capacity of enacting a position in relation to local/global tension. The users that I interviewed vacillated between two tendencies: to reject recommendations when they felt these were too "annoying," or to accept them when they felt these allowed them to be a part of conversations about music and technology. These two positions point to somewhat different cultural directions: whereas the first one emphasizes the need to make algorithms fit into prevailing notions of friendship and social behavior, the second one suggests that algorithms did not operate differently in Costa Rica when compared to the global north.

When users signaled their adherence to local culture, they also expected algorithms to fit within Costa Rican interaction traditions and rituals. As noted in chapters 1 and 2, Costa Rica is a country where "the importance of the collective and maintenance of harmony are valued over personal satisfaction" (Rodríguez-Arauz et al. 2013, 49). Drawing on this cultural notion as a resource, users demanded that algorithmic recommendations complied with local rules of public behavior. Accordingly, users expected that algorithms would hide their "face" rather than draw attention to themselves by providing one too many unsolicited recommendations that would disrupt harmony.

Alternatively, when people wanted to participate in global relationships, they adopted a discourse that emphasized ideas of individualization and quantification to assess algorithmic recommendations. This discourse reproduced the main tenets of how various platforms promote their services and algorithms (Prey 2018). Harvey captures the essence of this discourse with precision:

> Streaming platforms aim to zero in on the tastes of the individual listener. [. . .] the recording industry is [. . .] betting on a future of distribution and discovery dictated by quantification [. . .] to execute the recording industry's century-long mission: suggesting with mathematical detail what a listener wants to hear before they know they want to hear it. (Harvey 2014)

By adopting this logic, users suggested that algorithms in Costa Rica were not necessarily different from those used in other countries and strategically established equivalences among the operations of algorithmic platforms around the world. In this way, users suggested that, despite the geographic distances, they could be a part of the same global conversations.

Participants thus expected Spotify to function the same way everywhere. Explaining how he became a Spotify user, one interviewee recalled, "A friend who lives abroad said to me: 'This is what everyone is using now and you have to use it!' He was referring to using Spotify on the computer and on the phone." In his seminal work *Trust in Numbers*, T. Porter (1995) famously argued that quantification is a technology of "distance" that allows seeing phenomena from afar. Yet users in Costa Rica tended to envision the quantification embodied in algorithmic recommendations for the opposite reason: they valued algorithms as a technology of "proximity" that helped them feel connected to global conversations about music and technology.

INTEGRATING RESOURCES FOR RESOLVING SITUATIONS

In short, I have argued that Spotify users in Costa Rica followed algorithmic recommendations insofar as these allowed them "to construct, maintain, and refashion the 'cultured capacities' that constitute actors' basic repertoires for action" (Swidler 2001a, 71). This approach turns choice into a cultural process. Paraphrasing Swidler (2001a, 30), what made an algorithmic recommendation acceptable for my interviewees was how much it fit with particular cultural expectations, practices, traditions, and modes of life.

The integration of capacities that I have discussed in this section rests on how users thought that different cultural resources could help them resolve specific situations. On some occasions, such as discovering music that could be recommended to people whose opinions they valued, people deemed algorithms as an appropriate resource. Some of these situations were of a practical nature and required an immediate response. Yet other situations were moments or transitions in the life of a person. When I interviewed Rubén, the social psychologist I cited previously, he explained how his acceptance of certain recommendations could shift based on the situation he was experiencing: "I am changing now, at this moment you're interviewing me. If you had interviewed me four years ago, I would have told you: 'Mae [man], I'm only listening to things that are like Joy Division.' It was the only thing I wanted to hear and I spent like three years listening to that." In Rubén's view, this situation created conditions that made certain algorithmic recommendations more appropriate than others. Accordingly, he took a proactive role in obtaining these kinds of recommendations, which he did not necessarily feel the need to do at the time of our conversation.

For other situations, users discarded algorithms as valuable and opted for other cultural resources. Enrique, a 22-year-old public administrator, aptly captured this preference by indicating that, when it came to following recommendations, he applied a "manual filter" rather than algorithms. By this, he meant cultural sources that he thought contrasted with the notions of automation associated with datafication. He explained that, for him, radio seemed a much more appropriate way to get recommendations because he was familiar with the kinds of suggestions he could expect from each station. The *identity* of radio stations or programs offered him much more assurance than algorithms could, even if he didn't know which song would follow. In this perspective, habitus outweighed algorithms in terms of the certainty required for resolving certain situations and securing specific capacities (cf Airoldi 2022).

INTEGRATING RELATIONS ACROSS PLATFORMS THROUGH TIKTOK

A third type of cultural integration becomes obvious when examining Costa Rican users' relationship to TikTok's algorithms as part of larger assemblages of media technologies. A considerable number of studies over the past years have emphasized the need to further situate what people do with the media within complex webs of practices and technologies. In short, scholars argue that media studies have held a rather simplistic view of people's relationship with media technologies by isolating their practices with one platform at a time rather than exploring the interconnections individuals establish with and across them. Boczkowski and Mitchelstein (2021) thus speak of a "digital environment" constituted by a multiplicity of media technologies that is interconnected with natural and urban life. Similarly, Madianou and Miller (2013, 3) posited the notion of "polymedia" to stress how "most people use a constellation of different media as an integrated environment in which each medium finds its niche in relation to the others." Like the notion of "digital environment," people's relationship to "polymedia" must be situated within specific contexts. In a similar manner, Couldry (2016) coined the term "the media manifold," which he variously defined as "a complex web of delivery platforms" (31) or "the many-dimensional complexity of what it is we are 'in' with media" (38).

In this section, I build on these insights to consider how people integrate their experiences with a plurality of platforms into their algorithmic choices. I argue not only that people use various media technologies but also that they establish connections across them in ways that shape their understanding of, and relationship with, algorithms (Espinoza-Rojas et al. 2022). Algorithms are thus relational entities through which users of digital platforms mobilize a set of meanings and practices that shape their relationship with digital environments. To make this case, I emphasize how users constantly compared the algorithms of multiple media technologies to make sense of their relationship with TikTok and vice versa.

Scholars from various fields have stressed the importance of comparison in meaning making and learning (Gentner 2010; Higgins 2017). Comparison shapes how people learn about the features of objects and the information they learn about those objects (Higgins 2017). Comparison also leads to the discovery of relationships, similarities, and differences between objects (Gentner and Medina 1997; Goldstone, Day, and Son 2010; Hammer et al. 2008). TikTok users engaged constantly in comparative work. The focus groups I conducted lent themselves to this comparative endeavor in that they were designed to foster conversations about participants' opinions and to reach certain kinds of consensus (Cyr 2019). Yet the frequency with which participants compared TikTok to other platforms suggests that bridging and interconnecting experiences were central to the establishment of the app's identity in Costa Rica.

In what follows, I show the operation of four specific relational strategies through which users appropriated TikTok's algorithms: differentiation, aggregation, retrieval, and export.

DIFFERENTIATION

One key strategy was to differentiate TikTok from other platforms and their algorithmic operations. This strategy usually centered on the specific features that distinguished TikTok from other apps. Scholars in cognitive science and education studies have theorized two kinds of differences: alignable and nonalignable. According to Higgins (2017, 53), "An alignable difference is a feature-value difference between two items that occurs within a feature shared across both items. [. . .] A nonalignable difference is a difference that does not have a corresponding feature across the two items." For the most part, users envisioned algorithms as an alignable difference but other features in platforms' interfaces as nonalignable differences.

The preferred example to illustrate these differences with TikTok was Instagram. According to users, both platforms had algorithms, but their operation varied based primarily on each app's interface design. Rodrigo, a library scientist, neatly illustrated this idea:

> I attribute a large percentage of TikTok's effectiveness to its interface, because it has a scroll feature that is unique. It is unlike Instagram, which is more gradual. [On TikTok,] you have to do the [swipe] gesture to get a video. There's no text, everything is video, video, video. And that's the infinite scrolling.

For Rodrigo, differentiating TikTok from other apps led to a conviction that the app was "unique," not so much in terms of the content displayed but rather in how it functioned as a technological artifact.

Differences between TikTok and other platforms served to justify specific user practices. People offered one common example to signal how they tied the differences between these apps to various "shares of decisions"—as one person put it—as users. For many, the alignable differences and most significant distinction between TikTok and Instagram was how the "For You" and the "Explore" pages worked in the respective platforms. The users interpreted the design of both features as an invitation to do different things with these platforms. In their views, Instagram wanted users to follow certain people and pages by setting this as the default feature in the app. In contrast, they argued, TikTok fostered a sense of expectation in the "For You" page, because algorithms chose content from accounts the users did not follow. Accordingly, users felt they had to be prepared to receive unexpected content on TikTok as opposed to the predictability they associated with following selected individuals on Instagram. For many, this changed the overall motivation to use the app: by privileging the "For You" page, TikTok enabled them to forsake the obsession with following others and obtaining followers that prevailed in their use of Instagram and many other apps.

Differentiating TikTok from other platforms also had consequences for people's relationship with algorithms. For example, Laura, a college student, explained what she thought set TikTok apart from other apps: "I compare TikTok with Instagram's 'Explore' page. I suppose it's the equivalent of TikTok's 'For You' page. I feel like that's where you get to see TikTok's algorithm, you realize that [it] has studied you and your tastes way too well." For Laura, distinguishing TikTok and Instagram led to a better understanding of the apps' algorithms. This was a common conclusion that users derived from their comparison between TikTok and Instagram. Many users thus felt that, compared to other apps, TikTok's algorithms were more efficient and faster. This conviction strengthened their attachment to the platform (see chapter 2). It was because TikTok's algorithms seemed "unique" that users felt they needed to take care of it.

The sense that platforms were unique in certain ways also informed the notion that each app occupied a niche position in "polymedia" environments (Madianou and Miller 2013). This solidified the view that users lived in a differentiated ecology or digital environment with apps that played distinct roles and allowed these users to fulfill various communicative needs. Valeria, who graduated from the field of advertising but is currently unemployed, illustrated how users incorporated algorithms into their understanding of a differentiated media ecology:

> I am very aware of algorithms. If I see a TikTok [video] that I like on "Insta," I go to the account [of the video's creator on Instagram], look for the "watermark," write down the account or memorize it, then go to TikTok and follow the account or "like" the post or whatever so that [TikTok] keeps recommending more [similar content].

Valeria's relationship with algorithms thus stemmed from the sense that she lived in a digital environment of interconnected apps. Accordingly, she suggested that learning how to navigate this environment would result in a better opportunity to receive the kind of content she liked.

AGGREGATION

Another common relational strategy was to stress the similarities between apps (including TikTok) rather than their differences. This strategy usually worked through a form of analogical thinking that centered on identifying "how aspects of one item correspond to aspects of another item" (Higgins 2017, 46–47). Comparing TikTok to other apps led users to identify what they perceived as a common structure and to downplay the differences regarding specific features. Users aggregated apps into one single, overarching category consisting of cases that, while different in some specific regards, shared one main structure. The "social media" moniker was often employed to label this aggregation of platforms.

Once they had assigned a structural category to TikTok, users then mobilized a series of practices and understandings typical of their use of other platforms in this category and applied them to this app. In this sense, people treated TikTok's algorithms as a relational category, that is, "a category whose membership is determined by a common relational structure" (Kokkonen 2017, 780), in this case "social media." Anaité, an 18-year-old woman who said she was interested in the aesthetics of TikTok videos, narrated how she came to understand the app's algorithms: "If you linger a little longer on a post on Facebook, then you start to see similar posts. That's how it is on Facebook, Instagram, and Twitter. So [it has to] be similar on TikTok." Because she thought that TikTok belonged to the same category of media technologies, Anaité transported a set of assumptions about how the app's algorithms behaved and how she should respond to them. Likewise, users incorporated a series of specific practices that characterized their appropriation of platforms such as YouTube into their relationship with TikTok. The premise behind such practices was that, if users did what they had done on YouTube, TikTok's algorithms would successfully identify their preferences for (and aversions to) specific kinds of content, just like YouTube's algorithms did.

Whereas users contrasted TikTok primarily with Instagram, they drew on a wider variety of cases to explain similarities. For example, users often mentioned Spotify to compare how they thought TikTok's algorithms specifically recommended sounds. Mario, a 23-year-old international relations major, thus narrated how he incorporated his experience with Spotify into his understanding of TikTok:

> A few months ago, I read that Spotify had a background process that considered literally a thousand categories for each song: whether the song is upbeat, has acoustic guitar, uses minor notes or major notes, in order to personalize people's playlists or "radios." It seems to me that it would make sense for TikTok to work in the same way.

Mario's account starts from the assumption that, since both apps rely on algorithms to recommend content, it would "make sense" if their algorithms behaved in the same way.

Similarly, users employed the case of Twitter to ascertain how TikTok treated hashtags in its recommendations or relied on the example of Netflix to develop their understanding of how TikTok incorporated similarities in content to recommend specific videos. One user even suggested that Tik-Tok and Uber Eats were similar in how they provided "instant satisfactions." In short, users aggregated their experiences with multiple platforms into a general imaginary or stock of beliefs from which they drew to interpret all algorithms that belonged to that category (cf Bucher 2018). This aggregation strategy also worked to normalize processes such as surveillance. People suggested that, since other members of the "social media" family monitored users, it was "normal" that TikTok did it as well.

Part of the process of aggregating TikTok into a category of technologies was to project or imagine a future for it. Said Mario, "I really see TikTok as the next great social media platform. Many come and go, but TikTok for me is the next Instagram, the next Twitter." This narrative formalized an expectation around ideals of success. It conceived of a future where TikTok would reproduce patterns of development set by other members of its class.

RETRIEVAL

In his famous "Laws of the Media," Marshall McLuhan (1977, 175) argued that all media "retrieve [media] that had been obsolesced earlier." McLuhan envisioned the "law" of media retrieval operating at various levels. At a metaphorical level, he saw it as the recuperation of ways of understanding the media, as "'meaning' via replay in another mode" (McLuhan 1977, 177). McLuhan also emphasized how retrieval could work to revive past experiences, practices, and gestures; to "transfer power" between things (McLuhan 1977, 177); to reenact social ways of organizing around the media; and to perform identities in relation to the media.

Retrieval was key in users' relationship with TikTok and its algorithms. This was nowhere clearer than in how people considered the relationship between TikTok and apps that had previously existed and had now disappeared, most notably Vine. It was common for interviewees to capture this notion by suggesting that TikTok had what they called a "Vine energy." This comparison rested on what users thought were similarities in the kind of content available on the platforms. Through this retrieval dynamic, users integrated a cultural expectation into their understanding of algorithms. In other words, they evaluated TikTok by assessing whether its algorithms allowed them to reproduce their experience of Vine. One person expressed this with precision during a focus group: "I think of [TikTok] as this [thing that] has 'Vine energy.' I really couldn't think of any other way to categorize [it]." It was thus through familiarity with Vine that this person made sense of the experience of using a new app.

Some people decided to install TikTok on their cellphones precisely because it seemed like an opportunity to revive their relationship with Vine. Valeria thus recalled: "I realized TikTok had 'Vine energy' and thought: 'Say no more! I'm going straight to the AppStore!' I downloaded it and everything went very well. I consume it a lot." TikTok inherited this person's fondness for Vine as it met her expectations. But when the promise of retrieval was not satisfied, many users tended to evaluate TikTok negatively. For example, Nicolás (22 years old) indicated that the app's recommendations failed to capture the precision with which Vine showed him interesting videos.

In addition to considering TikTok as an iteration of previous apps, users also interpreted it as a means to retrieve essential aspects of an entire digital environment. This was the case when users emphasized how TikTok had been able to remediate or give a new incarnation to memes. Isabel was among those who expressed this idea with more precision:

> Sometimes I am chatting or talking with friends and I mention a TikTok [video] or a TikTok trend in the conversation. It's about virality. It has the value that the meme has. On Twitter, I see [TikTok videos] as forms of reaction. And I feel like the compilations of [TikTok videos] on YouTube are more curated content, just like Vine compilations were. They are like pure dank meme content, weird memes, that you are going to like.

As with cases discussed previously, Isabel mobilized the premise that TikTok's recommendations could only be assessed as part of a larger assemblage of technologies from which they derived their meaning. In her view, TikTok was an effective part of the "media manifold" in that it enabled her to materialize the practice of responding to others through memes to generate an emotional reaction. She then mentioned the case of YouTube to provide yet another illustration of how TikTok videos could retrieve collections of Vine videos.

Retrieval doesn't always culminate in the reproduction of past experiences in an exact manner. Identifying differences in how TikTok extended other practices and technologies often led to refining its identity or the sense that it was "unique" in some way. Lucía, a 21-year-old law student, thus said that TikTok not only remediated Vine but also in many ways surpassed it, because it allowed her to go beyond the focus on a single genre (comedy videos) and incorporated various kinds of content. Lucía's account centered on content diversity but also on the certainty that TikTok's algorithms could reliably reflect the interests of users. In that regard, she thought TikTok was insuperable.

EXPORT

Finally, users also suggested that not only could TikTok be interpreted in light of other apps but also that these other platforms could be reexamined through the lens of TikTok's design and identity. On such occasions, the direction of influence was reversed: TikTok was not the recipient of the meaning associated with other apps; instead, users "exported" TikTok cues to reinterpret larger digital environments. Through this dynamic, users assessed the identity and features of established apps against the model set by TikTok.

Exporting meanings to reframe the identities and functions of other technologies is frequent in the design and development of social media. The trajectory of "stories" is a case in point. Drawing on the work of Agamben (2009), I theorized this process as the constitution of "paradigmatic technologies" (Siles 2012a; 2017). Agamben identified the role of paradigms as making a singular context or phenomenon intelligible. In his words, a paradigm is "a singular object that, standing equally for all others of the same class, defines the intelligibility of the group of which it is a part and which, at the same time, [it] constitutes" (Agamben 2009, 17). Applied to the media, paradigmatic technologies are cases that make intelligible singular aspects of a whole set of artifacts. This notion thus calls into question taken-for-granted assumptions about the development of social media as a matter of gradual progress (captured by concepts such as the once-popular term "Web 2.0" or, more recently, "Web3").

For the users I interviewed, TikTok was a paradigmatic technology that made it possible to identify specific technological features in digital environments. For Mario, this became obvious when he realized that Instagram had begun to "copy" TikTok's most distinctive feature: algorithmic recommendations that seemed almost invisible but were so precise that they made him want to keep scrolling down endlessly. He explained:

> I noticed that Instagram already incorporated that same feature [on] its Home page. [On Instagram] you scroll the content of the people you know. One day, I realized that I was scrolling through content of pages that I do not follow [on Instagram], and I realized that it was because [Instagram] had already incorporated this feature of seamlessly recommending content to you. It seems to me that TikTok has set a very important trend.

Mario's use of the word "seamlessly" points to how he felt that Instagram was trying to emulate not only TikTok features but also the experiences associated with using those features.

TikTok was considered to be the source of specific usage practices that could be exported to other apps. Nicolás thus acknowledged that Instagram's incorporation of features available on TikTok had made him want to spend more time using Instagram. Finally, another telling example of TikTok's paradigmatic character can be found in interpretations of its influence not only on technology but also on culture in general. Users found evidence that TikTok had begun to shape culture because of its pervasiveness and the spread of content available on the platform. Tania, a public relations specialist, elaborated:

> You see something and associate it with [TikTok], or you listen to a song and it's like "*Mae*, it's that TikTok song!" I feel that TikTok is everywhere! There are even phrases or inside jokes. It's about how people understand them. And if someone doesn't understand them, it's like, "How do you not understand them?"

In this account, TikTok provides cultural opportunities for social interaction that must be recognized to avoid the possibility of being excluded from conversations.

In a similar manner, Yamila, a biotech engineer, noticed how she had started to incorporate content specifically associated with TikTok into her vocabulary. In her account, it was obvious that TikTok had shaped culture in ways that Instagram and Twitter had not. Both Tania's and Yamila's observations about TikTok's general influence on culture can be interpreted as the sense that TikTok was dominating their attention in their digital environments.

The case of TikTok points to the centrality of further situating people's relationships with apps and algorithms within larger assemblages or ecologies of digital media. Living in the "media manifold" provides users with opportunities to incorporate specific relations across platforms—such as to differentiate, aggregate, retrieve, and transfer meaning and practices—into the repertoire of cultural elements that shape their ties to algorithms.

CONCLUDING REMARKS

In this chapter, I have stressed that choice does not occur in a cultural vacuum. Instead, algorithmic recommendations need to be culturally enacted (Seaver 2017). To this end, people integrate algorithms into the repertoire of cultural resources that allow them to resolve and respond to different situations and demands in their lives. By using the notion of integration, this chapter has emphasized how culture matters in people's choices much more than what has been recognized in the scholarly literature on datafication.

Integration characterized how the users that I studied enacted algorithms on all three platforms. In this chapter, I also showed the existence of differences in the modalities of how integration works. These differences can be explained as an outcome of the specific situations that users wanted to resolve in their daily lives. In the case of Netflix, I focused on how users relied on different *sources* to decide what recommendations to follow. For Spotify, I privileged an account of how they sought to acquire *capacities* by privileging certain music choices and technological features. Discussing the example of TikTok, I revealed how users also engage in comparative work to establish numerous *relationships* across media technologies in their digital environments. Combined, these sources, capacities, and relationships constitute a cultural repertoire from which the users can draw. This view is consistent with Swidler's (2001a) definition of culture. For Swidler (2001a, 40), "[People] do not simply express perspectives or values instilled in them by their culture. Instead, they draw from a multiform repertoire of meanings to frame and reframe experience in an open-ended way." What I have called "integration" is the process of enacting this repertoire by choosing the resources that best correspond to the situations that people face.

This chapter has also sought to offer a more symmetrical account of choice. As Couldry (2016, 27) notes, what people leave out of their choices has been "deeply neglected in the first 30 years of media studies but has become ever more essential, both analytically and practically" (27). In this chapter, I have considered the factors that shape both instances of acceptance and of rejection of algorithmic recommendations.

My analysis of Netflix focused on the construction of cultural repertoires that people referred to when deciding on what to watch. Integration here designates the act of variously combining different resources under conditions shaped by people's social, cultural, personal, and professional codes. It is against the backdrop of each person's sociocultural position that algorithmic recommendations acquire meaning as relevant resources.

Considering the case of Spotify revealed how people constantly engage in integration work to acquire specific capacities. As with Netflix, Spotify's algorithms are only one among many cultural resources that users can draw on to enact their choices. While people might deem algorithms appropriate for some situations, other occasions present a distinct set of challenges that need to be resolved differently. These situations might involve short-term activities, such as choosing the appropriate kind of music in order to study or do homework. But they can also refer to periods in the life of a person. Spotify users drew on the different resources from their repertoire of cultural tools—both algorithmic and "manual" filtering, as one person put it—to resolve those situations.

As in chapter 2, the local–global tension was key in how the Costa Ricans that I studied enacted algorithms. For some, it was crucial that algorithms complied with local rules of public behavior. Accordingly, they assessed the pertinence of algorithmic recommendations based on the extent to which these could compromise harmony and social order, which are highly valued in Costa Rican society. But on other occasions, people enacted algorithms to participate in global conversations about music and technology. On such occasions, they abandoned more local ideas and reproduced instead cultural discourses that surrounded algorithmic platforms in places like the United States. Algorithms thus allowed them to channel the aspiration of cultural proximity with the global north.

Finally, my discussion of TikTok revealed that integration takes place in the material conditions created by specific media environments. Combined with the study of Netflix and Spotify, this analysis showed the need to situate the study of algorithmic choice within the conditions created by platforms' embedding in the structure of everyday life *and* the inter-relationships that characterize today's digital environments. In this way, it becomes possible to account for people's relationship with algorithms in ways that recognize people's histories with previous artifacts, their ties to a plurality of contemporary platforms, and the construction of an ecology of technologies in which people think that each app has meaning and each practice a place. How users ritually formalize these meanings and practices into patterned actions is the topic of chapter 4.

Rothenbuhler (1998, 27) defines rituals as "the voluntary performance of appropriately patterned behavior to symbolically effect or participate in the serious life." By "the serious life," Rothenbuhler referred to those issues that pertain to the domain of what people consider to be important in their lives for a variety of reasons. In this chapter, I argue that users' relationship with algorithms is organized in formalized practices that take place at specific times and places, and that these practices can be identified as rituals.

This argument is supported by Carey's (1992) approach to communication, in which he also foregrounded the notion of ritual. Carey employed it to stress the constitutive character of communication practices and exchanges. As Rothenbuhler (2006, 14) summarizes it, rituals are "consequential": they "construct [. . .] the realities in which we live." Ritual has become a keyword in the study of media (both "old" and "new") (Burgess, Mitchell, and Muench 2019; Cui 2019; Rothenbuhler 1993; 1998; Silverstone 1994). In his analysis of communication studies as a field, Craig (1999) even situated ritual (and the constitutive approach to communication it expresses) as an intellectual touchstone that could allow the field's different theoretical traditions to take part in a dialogue.

In this chapter, I analyze the significance of rituals in relation to algorithmic platforms. I specifically draw on Couldry's (2003, 2) conception of rituals as "actions organised around key media-related categories and boundaries, whose performance [. . .] helps legitimate the underlying 'value' expressed in the idea that the media is our access point to our social centre." Couldry's approach turns rituals into a key mechanism for exploring issues of power. According to Couldry (2009), rituals naturalize order (rather than express it) by formalizing categories and boundaries through performances that sustain their legitimacy. In short, rituals underlie what Couldry (2003; 2012) calls "the myth of the mediated centre," the notion that the media are indispensable in bringing people together and are the privileged entry point to understand what societies value, think, and do. Over time, replaying this myth through the performance of rituals has normalized the media's concentration of symbolic power (Dayan and Katz 1992).

More recently, Couldry (2015) warned about the rise of a similar myth in the case of social media which works to legitimize the view that these platforms bring people together. He called this the "myth of us." In Couldry's (2015, 620) words:

> A new myth is emerging about the types of collectivity that we form when we use social networking platforms: *a myth of natural collectivity* whose paradigmatic form lies in how we gather on platforms [. . .] This myth offers a story focused entirely on what "we" do when, as humans like to, we keep in touch with each other. [Emphasis in original]

This chapter builds on these insights to discuss the emergence of a particular kind of myth that is enacted and reproduced through the ritual use of algorithmic platforms. This myth goes beyond the emergence of a particular notion of collectivity on social media. It also involves legitimizing the category that algorithmic platforms are the center around which people's practices, moods, and emotions gravitate. Thus, adapting Couldry's own definition, I refer to this as "the myth of the platformed center" or "the belief, or assumption, that there is a centre to the social world and that, in some sense, [algorithmic platforms] speak 'for' that centre" (Couldry 2009, 60). I argue that rituals around Netflix, Spotify, and TikTok normalize the idea (or category) that users need to cultivate practices, moods, and emotions, and that algorithmic platforms are the obligatory intermediary in this process. This creates a space in which algorithmic recommendations acquire particular meaning for users.

In this chapter, I examine ritualization dynamics in relation to these three platforms in Costa Rica. I begin by discussing the proliferation of rituals on Netflix and how the platform structures its recommendations around these rituals. This legitimizes the notion that algorithmic suggestions are a natural outgrowth of people's consumption practices. The chapter then focuses on how users respond to the exigencies of affect by ritually creating playlists on Spotify that allow users to cultivate moods and emotions. By doing so, rituals become the basis on which users evaluate the affective legitimacy of algorithms. Finally, I examine how Costa Rican users turn to TikTok to deal with boredom. The chapter considers boredom as a moral emotion that users feel the obligation to resolve and shows how algorithmic personalization is normalized as a solution to boredom that needs to be ritually enacted.

OF RITUALS AND RECOMMENDATIONS ON NETFLIX

The myth of the platformed center operates by connecting algorithms with users' ritual appropriation of digital media in various ways. In the case of Netflix, this is accomplished by establishing algorithmic recommendations as a direct and straightforward result of what users have watched previously on the platform. In this way, algorithms are positioned as an outgrowth of users' rituals. I begin this section by examining the types of rituals that characterize the appropriation of Netflix in Costa Rica. I then consider how rituals create the specific context in which Netflix offers algorithmic recommendations and how these acquire meaning for users.

TYPES OF NETFLIX RITUALS

Rituals on Netflix operate through a twofold process: they require users to adjust their daily live to accommodate watching content on the platform, as well as adapting the platform to work around daily life activities. In turn, based on the technologies users employ and the content they watch when they perform their rituals, Netflix makes specific recommendations to (further) shape these processes.

I have identified three types of domestication rituals on Netflix: individual, collective, and a hybrid of the two. The first kind of ritual centers on the individual use of the platform. Users typically made the consumption of Netflix compatible with the most mundane activities. As interviewees put it, Netflix was consumed "when I eat," "when I wash clothes," or "when I'm getting ready." Building on the personalization dynamics examined in chapter 2, the Netflix subject functioned as an Other, a companion during these activities. This is not unlike what audiences have previously done with television or other media. Jimena, a 29-year-old freelance consultant for small organizations, explained: "I sit down for lunch and turn Netflix on. [I put] some show on, like *Friends*, even if I have seen it a thousand times. I do it to have a companion. When I iron my clothes, I turn on Netflix." Netflix makes it easy to perform this ritual: it can be used on different devices, at any location, and it can be interrupted and resumed without major alterations. Because Netflix can be so easily incorporated into the spatial and temporal conditions of daily life, watching content on the platform and carrying out daily chores need not be separated.

By performing the ritual individually, users felt they could distribute their attention between various activities as they pleased. Yet engaging with the platform holds a distinct space in the sequence of activities that formed the ritual. Lucrecia, a 24-year-old nurse, explained:

> This is my ritual to go to sleep: I put on my pajamas, brush my teeth, remove my makeup, and get ready. When I'm already in bed, I turn on the computer. I try to be alone [and] in my bedroom, because that indicates privacy. I watch [Netflix] at night and I don't like to be interrupted when I'm watching it.

At the time I interviewed her, she had been watching *Spartacus*, a series she selected because her boyfriend did not like it. For this reason, she could perform this ritual alone. For Lucrecia, the bedroom offered a space of relative independence from others. Watching Netflix put an end to the sequence of patterned actions that defined the conclusion of her day.

This ritual also required users to adjust to the platform to provide a better fit for the experience. Users looked for specific content, such as shows, given that they are episodic (which helped the repetitive nature of the ritual) and are shorter than movies (and thus made it easier to allocate time throughout the day). Although this kind of content was the most commonly watched among interviewees, individual rituals were not restricted to (short) episodes. Users also reported watching movies across various sessions throughout a day or week. As Lucrecia's example showed, personal devices that are easier to carry around, such as cell phones, tablets, or personal computers, were usually preferred to perform this ritual. These devices also helped users to extend their use of Netflix outside the household to the office, their education centers, or on public transportation (cf Hartmann 2013; 2020).

Another kind of ritual characterized the collective domestication of the platform. As with individual consumption, adjustments were required to daily life, content, and technology in order to carry out this ritual. As the name suggests, collective rituals involved others. They also generally occurred at preestablished times. These rituals relied on the establishment of boundaries: a form of social interaction was created around Netflix by deciding who got to participate in them. In this sense, these rituals mattered in that they represented a means to define ways of being social.

The purpose of collective rituals transcended entertainment and focused on interpersonal connections with members of the family, partners, or friends. These practices fed on the cultural principle, well established in Costa Rican society, that the best way to cultivate a relationship is to spend time together. Mónica, a 24-year-old woman who works in marketing, described the most defining ritual in her life: "It's the Saturday movies that I watch with my mom and my sister. I have to look for something that all of us like, which is a bit [*un toque*] difficult." Many interviewees were aware of the significance of the ritual as a form of bonding and thus carried it out in a reflexive manner. That was the case of Luna, a 28-year-old public health specialist, who lives with her mother and her mother's boyfriend. She watched Netflix at preestablished times to bring members of her household together. In her words, "We do have shows that we watch together. We have a ritual for it, and that's part of our dynamic." At the time of our interview, the section "Continue watching" on her profile included mostly shows branded as "Netflix Originals," such as *Queer Eye*, *The Queen of Flow*, and *RuPaul's Drag Race*. Luna's words are revealing in that she acknowledged the role of rituals in turning her home into a family through Netflix (Silverstone 1994).

Users carried out such practices with different members of a household to pursue specific ties with them. That was the case of Mariana, a 49-year-old lawyer, who said she had various rituals that differed based on with whom she was watching Netflix. She explained that she watched content about fashion or "reality shows" with her daughter, movies for kids with her youngest son, and contemporary shows or dramas with her eldest son. Mariana said that, although she clearly had a personal preference for romantic movies, she was willing to watch other kinds of content that would allow her to spend time with family members.

For the most part, users turned to television (the artifact) to perform this ritual because the screen is larger and has better sound than small devices have. The most common content for this ritual were movies (which have a specific duration and narrative structures that help bind the ritual temporally), but users also selected shows and watched (or binge watched) several episodes.

Finally, I also identified another type of practice that bridged individual and collective rituals. Users indicated that they watched certain shows simultaneously (mostly with friends) at a distance (even before the COVID-19 pandemic). Thus, this ritual is individual (because the user is alone when watching the content) but also collective (because someone else is watching simultaneously somewhere else). Interviewees employed terms such as "Netflix party" to name these practices, referring to the browser extension that allowed them to watch the platform remotely with other people. Carla, a college student, explained:

> We don't share a physical space to watch [content], but we watch shows at the same time. For example, along with a girl friend, we started watching *Gilmore Girls* again. We kind of said, "Let's watch two episodes today." We both went to our [own] house[s], did our part, and talked about it [as we watched it.]

Conversations about content watched with others can take place synchronously or asynchronously, both through messaging apps and in person.

The prevalence of rituals reveals the continued relevance of Silverstone's work. The processes of objectification and incorporation remained highly relevant for my interviewees: domesticating Netflix was performed at certain times of the day and required locating certain technologies in specific places. It was precisely through these rituals that users normalized the notion of Netflix as a social center. In other words, through individual, collective, and mixed rituals, users acted out the centrality of Netflix in their daily lives.

"BECAUSE YOU WATCHED"

As Kant (2020, 34) notes, algorithmic recommendation

> is premised on the idea that your *future preferences* can be inferred from your *past interactions*. Your previous click-throughs, likes, and browsing histories are assumed to be the best means of computationally inferring your present behaviors, and so your past [. . .] trajectory goes on to determine what you will see next by. [Emphasis in original]

Following this principle, Netflix employs several mechanisms to connect algorithmic recommendations to these rituals. For example, the company tracks "the device that the member is using" and "the time of day and the day of week" when rituals take place (Chandrashekar et al. 2017). In this way, Netflix can adjust recommendations to suggest certain kinds of content.

Rituals create the context in which algorithmic recommendations acquire meaning for users. In other words, users make sense of their "past interactions" with Netflix (to use Kant's phrase) not as loose practices but rather as specific rituals. It was thus common for interviewees to trace what they thought were the origins of Netflix suggestions back to specific rituals they performed. That was the case of Elisa, a 39-year-old photographer, who noted:

> On the weekends, I watch Netflix with my nephews. And every Sunday, I watch it with my mom. In fact, now that I check my Netflix account with you, you can see that there are two patterns: one is the shows that I watch with them, and the other one is my series.

In Elisa's account, every algorithmic suggestion had an intelligible origin that justified its presence on her profile. The use of the term "pattern" also indicates that Elisa thought that Netflix could identify the practices she repeated regularly and provide suggestions accordingly.

By the same token, users assessed Netflix's recommendations based on their potential to carry out rituals. For example, when asked to explain how she chose content for those preestablished moments when she watched Netflix with her mother, Carla noted that she typically pondered whether the movie's name and overall description were consistent with previous content they had watched together. The two most recent films they had watched together on the platform were *Isn't It Romantic* and *Bad Moms*. In a similar manner, Cleo, a 20-year-old psychology student, explained how her collective ritual with her boyfriend typically unfolded:

> We start by saying, "So, what do we want to watch?" We then look at what Netflix recommends. If nothing catches our attention, we then say "OK, let's just watch *Friends*," to play it safe, or *Phineas and Ferb*, which we both like.

In this example, algorithmic recommendations were welcomed, because they allowed a couple to perform the ritual of spending time together and having something to talk about. When algorithms didn't interpellate them, they resorted to shows that they had already watched, just to be sure the ritual would evolve without problems.

As noted in chapter 2, Netflix also relies on generic cues to enable users' assessment of algorithmic suggestions. Users typically employed genres recommended by Netflix as an instrument of consensus in collective rituals. Mónica, the marketing specialist, explained that she typically examined main genre categories one by one to decide what she wanted to watch with her mother. Like Carla and Cleo, Mónica argued that she evaluated algorithmic recommendations based on their potential to perform rituals. When these recommendations failed, she admitted she typically watched something else on the platform and rarely looked for other alternatives. In this way, Netflix retained its centrality in these users' lives, and the ritual worked to maintain the myth of the platformed center.

Consistent with the principle of algorithmic prediction summarized by Kant (2020) above, a key strategy employed by Netflix to shape users' rituals is to suggest that recommendations are a direct result of their past behavior and, therefore, that it is their responsibility to care for them. This strategy works by connecting recommendations to consumption rituals in the lives of users. Building on the ideas of Althusser (2014), Fiske (1992, 216) noted that "the most significant feature of [ideological apparatuses] is that they all present themselves as socially neutral, as not favoring one particular class over any other." In a similar manner, Netflix presents itself as neutral by suggesting that, through algorithmic recommendations, it merely reflects previous user actions and behavior patterns. This is accomplished with Netflix's iconic category "Because You Watched . . .". Figure 4.1 shows an algorithmic bundle displayed on Marcia's profile. It associated the consumption of Netflix's romantic movie *Love Is Blind* with notions of dating, traditional wedding rituals, wealth, Latina womanhood, and kindness. Moreover, this bundle also articulated the ritual of watching a movie and a series of cultural products.

When algorithmic recommendations and bundles resonated with their rituals, the users that I interviewed tended to accept them as a natural consequence of their consumption practices. During the interviews, a typical way to explain why users thought they had received certain recommendations was to say, as one person did, "It's because I've led [it] to this." The acceptance of algorithmic recommendations came from the conviction that users could locate the source and principles of the suggestions they received. In other words, the more users thought that they knew why and how Netflix was making recommendations to them, the more receptive they were to such recommendations. This conviction came from the experience of using the platform ritually and noticing certain patterns in Netflix's operation. For example, Elena, who is 24 years old and works for a nongovernmental organization, said she willfully put herself in a position to receive recommendations because she was convinced that Netflix "knew" what would be of interest to her to carry out her rituals.

The users that I interviewed normalized the operation of algorithmic bundles and maintained that Netflix recommended content to them by establishing patterns of association between what they had already watched on the platform and "similar" kinds of content. For example, Cleo, the psychology student, stated: "I suppose [Netflix] has something like a system: if

Figure 4.1
Algorithmic bundle derived from content previously watched by Marcia, taken from Netflix's "Because You Watched" recommendation feature on the user profile page.

I saw this show, it will recommend similar things to me, based on the genre or the plot." Users thus believed that the Netflix Subject "knew" with certainty what users had done on the platform and recommended content that reproduced generic patterns with mechanical predictability.

Users also explained the "similarity" between content previously watched on Netflix and algorithmic recommendations they received as resulting from more technical or formal issues. Carolina, a public relations specialist, thought that Netflix could link together media texts with certain kinds of "colors." For example, she spelled out why she thought Netflix was recommending the film *The Perfect Date* on her profile by indicating there was an aesthetic "pattern" with other movies she had enjoyed, most notably *To All the Boys I've Loved Before*. By using the term "pattern," she suggested that recommendations were not random but instead were calculations made for her based on her own actions and preferences. Other interviewees interpreted Netflix as establishing this pattern of similarity by taking into consideration how many movies with specific actors and actresses they had watched.

Despite their willingness to accept algorithmic suggestions that fit with their consumption rituals, users also questioned the notion that Netflix was only mirroring their previous practices. For example, they criticized Netflix's supposed neutrality by identifying temporal patterns behind the recommendations driven by commercial interests. Thus, Inés, who said she ritually watched romantic content, distanced herself from certain suggestions that were displayed on her profile during our conversation. On her profile, the first algorithmic bundle recommended to her was "*Programas románticos*" ("Romantic series"), which included several Netflix shows and films such as *Love is Blind* and *Morocco: Love in Times of War*. To explain this, she noted: "This [type of recommendation] is not permanent. It [Netflix] is showing it because Valentine's Day is close." She thus noted a commercial bias to promote certain content that did not derive precisely from her own practices but rather from an external cause. In a similar manner, interviewees strategically compartmentalized recommendations based on the rituals they sought to carry out. In this way, they downplayed the relevance of recommendations by showing that they were not appropriate all of the time and for all circumstances.

Netflix's case shows the myth of the platformed center in operation. By employing strategies such as naming categories in particular ways ("Because You Watched . . ."), Netflix establishes the idea that algorithms mirror the ritual practices of its users as a basis to legitimize certain predictions. Users then assess these recommendations by considering the potential they have for carrying out their individual and collective domestication of the platform.

AFFECT, RITUALS, AND PLAYLISTS ON SPOTIFY

Appropriating Spotify involves similar kinds of rituals to the ones described in the previous section. Users related to the platform through a series of formalized and patterned actions. But a focus on users' relationship with Spotify and its algorithms allows one to identify a particular kind of ritual process: the cultivation of affect. Scholars have already noted the importance of affect in people's relationship with music (DeNora 2000). In the case of Spotify, users have increasingly turned to playlists to materialize affect into an artifact and thus cultivate moods and emotions. On its support website, Spotify (2019b) promotes playlists as "collection[s] of music. You can make them for yourself, you can share them, and you can enjoy the millions of other playlists created by Spotify, artists, and fans." Dias, Gonçalves, and Fonseca (2017, 14379) define playlists as "ordered sequence[s] of songs meant to be listened to as a group."

In what follows, I examine the creation of playlists by users as a ritual that sustains the myth of the platformed center. I trace the evolution of this ritual from its origins in the perception that affect demands a response from users to the practices involved in materializing this exigency into a specific kind of artifact (i.e., the playlist). Then I consider how this ritual creates the conditions for establishing an affective relationship between users and algorithms that normalizes the primacy of Spotify in how moods and emotions need to be cultivated.

CULTIVATING MOODS AND EMOTIONS

Music, Hesmondhalgh (2013, 11) writes, "is a cultural form that has strong connections to emotions, feelings, and moods: the domain of affect." This aptly describes the relationship that Spotify users in Costa Rica had with music. They typically employed the platform to cultivate this affect in particular ways. I use the term "cultivation" to emphasize various issues. First, this term calls attention to the work involved in producing moods as outcomes. Capturing, eliciting, exploring, or maintaining moods and emotions requires significant time and involves specific practices. Second, cultivation stresses the ritualistic nature of this work. Obtaining moods and emotions with and through music is a constant endeavor, reproduced every day and repeated throughout the day. Third, cultivation brings to the fore the dynamic nature of users' relation to affect. It posits affect as both source and outcome; it situates affect and music as relative products of one another. For these reasons, I prefer the notion of cultivation to others traditionally used in social science research, such as mood management (Knobloch and Zillmann 2002; Zillmann 2000).

Most people that I interviewed used Spotify to produce specific moods and emotions through music. In the words of Mateo, a 19-year-old college student, music was for him "the engine of how you want to feel." To be in a mood thus means "to think and feel at length through that mood about everything" (P. A. Anderson 2015, 817). Users typically experienced the production of specific moods and emotions as a need that required a resolution. For example, Ricardo, a 31-year-old software developer, linked this need to the demands of his job:

> I do choose music based on how I want to feel, not so much the other way around. This is not only because I need to concentrate but also because I need to concentrate on a certain kind of job, so I put on *that* kind of music. [This helps to] enter the atmosphere we want for the product we are developing. [Emphasis added]

He illustrated this dynamic by putting on music created in the 1980s to inspire himself to work on a video game that had what he described as "an 80s look." Similar accounts were given for many other practices, most notably exercising and studying. The basic premise was that music allowed people to enter a particular kind of affective state where moods and emotions flourished in ways that could allow them to perform professional or personal tasks under the desired conditions.

Users turned to music to create not only individual but also collective moods and emotions. This was done both in work environments and domestic settings. Eugenia, a 22-year-old journalism student, narrated a common situation that involved other members of her household:

> I'm in charge of music at home; my brother supports it and my parents don't know how the sound equipment works. [I choose music] that helps to transport people from one place to another. That's what I try [to do]: to create an environment so [that everyone] is in tune, on the same page.

In this way, music was a key in producing a collective mood through a ritual performed at Eugenia's household. Mastery of technology—that is, the sound equipment—was what allowed her to control the ritual.

Another dimension of affect cultivation involved the use of music, not to create, but to respond to moods and emotions derived from specific experiences and activities. For Pia, a 29-year-old specialist in natural resources management, music provided a language to articulate affect. Some users even suggested that moods and emotions were incomplete without a musical expression. For Mateo, music is what allowed him to sustain moods and emotions over time. He planned his music selections accordingly. In his words: "If I listen to something, it's because I will be in that mood for some time." A common manifestation of these ideas was the use of music as a soundtrack for daily life activities. Music was typically seen as the perfect companion for "whatever I'm doing," as Enrique, a 22-year-old public administration student, put it (cf DeNora 2000).

Finally, users also turned to music to explore the temporality of their lives. In this way, music became a way to revisit and restore specific moods and emotions associated with events in the lives of users. Pia explained:

> There are songs I remember because I used to go to a bar with some friends and these songs were played there. When I hear them, I say, "That's from when I used to go to that bar!" They connect me with that part of my life.

To crystallize these moods and emotions, users performed the ritual of creating playlists.

THE AFFECTIVE ORIGINS OF PLAYLISTS

Playlists began with an affective exigence. In other words, users turned to playlists to produce, capture, and explore moods and emotions associated with a variety of events and experiences in their lives. A common source of exigence came from ordinary activities. Adrián, a 22-year-old public relations specialist, provided a telling example. During our conversation, he explained how a mundane activity demanded the creation of a new playlist: "This [playlist] is called 'Embers' (*Brasadas*). I was smoking with my friends and it was like 'I *have* to make a playlist for this,' and only songs along this line started to come to my mind" (emphasis added). In this way, Adrián sought to create an appropriate affective state for this ritual in his life.

Like Adrián, interviewees described many other mundane activities that became opportunities for creating playlists: playing a video game, reading a book, watching a movie or a series, studying, thinking of a person or concept, or simply having an idea. These situations operated as exigencies, in that users experienced the possibility to cultivate the moods and emotions linked to these situations as an objectified social need (Miller 1984). In this way, users began incorporating the playlist as an obligatory means to make sense of moods and emotions.

Another source of affective exigence came from extraordinary events. Leo, a 20-year-old audiovisual producer and musician, explains:

> The common trigger for all the playlists I have are very important moments in my life. On November 27 of last year, an online friend died. It was a very sad day so the first thing I did was to create this playlist. [He] was from Brazil so all songs are in Portuguese or are from Brazilian artists.

The experience demanded a response from the user, who envisioned the playlist as a means to produce an affective state that could work as a tribute to his friend. What counted as an extraordinary event was relative; during interviews, users mentioned examples that included the first day of college, Christmas, the end of the academic semester, the month before the end of the semester, a party, an evening out with friends, a particular class, or a concert attended or planned, among others.

Whether about mundane events or something extraordinary, playlists were all about capturing the proliferation of moods and emotions. Accordingly, for many users, playlists were never an individual or unique creation; they were one instance of a larger series. This is what Leo expressed when he claimed: "On Spotify, making a playlist is a ritual." Playlist creation is an affective reaction integrated into everyday situations. It is thus not surprising that, when asked to describe her playlists, Abril, a graduate student in linguistics, stated, "I've got everything in here." By this, she meant both that playlists responded to every experience one could think of and that all that was important to her was represented by the ensemble of playlists she had created over the years.

FROM EXIGENCIES TO ARTIFACTS

Historically, media technologies have afforded ways to materialize affective exigencies. Many interviewees envisioned CDs and mixtapes as antecedents to their playlists. The features afforded by Spotify to automate this process are what drove some of them to the platform in the first place. Nina, a 52-year-old audit specialist, explained the convenience offered by Spotify when compared to other technologies she had used in the past:

> You had to begin by classifying and identifying cassettes. Now you don't have those complications. [. . .] If you wanted to make a playlist, you needed to call a radio station and ask the person: "Could you play that song so I can record it?" [. . .] It used to take time. And you had to carry around cassettes and CDs. Now you just show up at a party with your phone and [ask]: "Does the speaker have bluetooth?"

Creating and maintaining a playlist is now codified in a few automated actions on Spotify's platform. A video available on Spotify's (2019b) support website instructed users on how to create playlists by following five steps: "1. Play song then tap. . . . 2. Tap add to playlist. 3. Tap create. 4. Enter playlist name. Tap create. 5. Tap [to] add more songs."

But building and maintaining playlists still takes time and effort. A key in these processes was reconciling the exigencies of affect with specific musical substance by establishing some sense of consistency (Lena 2012). Users looked for songs that seemed to them to be coherent with the mood and emotions intended and then "tapped to add" them to the playlist. Several interviewees called this process finding the playlist's "line"; others used terms such as "thematizing" it. By these terms, users referred to the search for a pattern of similarity through which they articulated songs that could work to produce, respond to, or explore specific moods and emotions.

The use of genres as a relatively predefined set of musical properties—or what Lena (2012) calls "styles"—was flexible. However, users drew on this notion of genres to establish the sense of similarity between songs that could be included in playlists. During the interviews, it was common to find references to playlists that had a "rock" or "jazz" feel. These definitions honored a preexisting symbolic contract and usually met the expectations about the kind of music they contained. But users were also quick to note the limitations that these labels had in defining music, playlists, and themselves. The most telling answer in this sense came from Rubén, a 39-year-old psychologist, who stated: "I am multi-genre." Javier, who is 20 years old, put it differently: "The 'genre' does not matter as much as what the song evokes in me." In this sense, playlists exceeded the boundaries of traditional genres, because affect afforded opportunities for finding new roles for music. These opportunities were not unlimited but rather were culturally and historically situated (Nowak 2016).

Selected songs were then organized to provoke the appropriate affective response. Ricardo, the software developer, explained his approach as follows: "What I do is to grab *similar* music and arrange a playlist; I do spend some time to put it in the order I like" (emphasis added). Users employed notions such as "rhythm" to describe the goal of this practice. The notion that users were engaging in storytelling through the song order was not uncommon (cf Dhaenens and Burgess 2019). Gabriela, a 21-year-old sound engineer, similarly maintained: "I have a concept, a story, or something I want to tell to somebody, so I do it through the songs [of the playlist]." To crystallize the affective exigencies imposed by these abstract "concepts," she experimented with the grammar of the playlist (both lyrically and sequentially). She defined the result as "narrative playlists."

Users also approached the name selection carefully and reflexively. Leo, the musician, compared the names of playlists to book titles. Naming playlists meant naming a chapter in the affective life of the user and was a way to bring the combination of songs to life. Although the platform automatically creates an image or thumbnail for each playlist, some users chose images themselves. Roberta (20 years old) asserted:

> I like to be as creative as I can because I believe music represents a lot of one's personality. I try to use [images that] are related to what I'm going to hear, but at the same time it's an issue of aesthetics. I do believe that [images] give each playlist its personality.

Playlists thus got a specific name and look to reflect the user's own personality. In this way, the playlist and the person become a reflection of each other.

The process of creating a playlist can be tied not only to antecedent media technologies, as noted above, but also to other cultural practices. For example, playlist creation holds evocative parallels with the practices of DJs. Like DJs, playlist creators take finished products and transform them into a performance (Greasley and Prior 2013, 25); they carefully design transitions and develop sequences of music "sets"; and they seek to bring potential audiences to particular affective states (M. Katz 2012). It would not be an overstatement to suggest that, through playlists, Spotify users turned themselves into the DJs of their everyday life.

THE ENDINGS AND BEGINNINGS OF PLAYLISTS

On some occasions, playlists found an ending. Sometimes they were purposefully finished; sometimes they were abandoned or ignored. The deliberate end of a playlist was usually met with emotion. Eugenia, the journalism student, recalled:

> I made this playlist for my birthday. I am very proud of it because I have been working on it [for a long time]. When I finished [it], it was 44 hours long. I am [so] proud that I wrote on Twitter: "Hey, I have a very cool (*tuanis*) playlist, if someone wants access to it, let me know."

Eugenia's words captured the emotional attachment that users had with their playlists: the lists were seen as valuable possessions. Spotify then helps transform these possessions into commodities in a market of moods and emotions.

The reasons for explaining why users stopped including songs on playlists ranged from practical (they take time and effort) to affective (new moods and emotions demand attention). They were also cabalistic: Leo, for example, claimed that his playlists were finished once they included forty-seven songs. More generally, the potential of playlists for affect cultivation was not unlimited. At some point, music failed to produce or capture the desired moods and emotions. Users typically employed Costa Rican slang to refer to this issue: songs and playlists get "burned" (*están quemadas*). This expression suggests that, each time a song or playlist was reproduced, its affective potential diminished until they were turned to ashes and couldn't ignite an emotional fire.

Not only do playlists have symbolic worth, but as noted previously, they were seen as a prime instance of self-expression. The fact that these playlists were hosted in users' profiles on the platform encourages this process. According to a person I interviewed: "Playlists [have to] be under my name, my username, because they are like my identity." For this reason, users almost never deleted playlists, even if abandoned or forgotten. Abril, the linguistics graduate student, captured a common reaction when I asked her whether she had deleted any playlist: "No, no, no, I never delete them! Because I say to myself: 'I might come back,' even if I never do." This assertion reveals how playlists came to play a practical role for Spotify users: they automated access to both the songs and the moods and emotions enacted by their genres.

As a ritual, the end of playlists was also a starting point. When one play-list was (relatively) finished, new opportunities arose to restart the process. Moreover, once they were concluded, the playlists became the place where the subsequent use of Spotify actually began. Many interviewees reported going directly to their playlists when they started using the platform. In this way, playlists provide a shortcut to the cultivation of affect.

AFFECT AND ALGORITHMS

As for the case of Netflix, rituals on Spotify created a context where algo-rithms acquired particular meaning for users. Specifically, building playlists to cultivate moods worked to infuse algorithms with affect. Accordingly, my interviewees said they welcomed algorithmic recommendations that could be included in a playlist. Spotify promoted this practice through several features on the platform, such as the "radio" that launches algorithmic recommendations associated with specific songs, albums, or artists. Con-sidering the possibility that these recommendations would allow them to further explore a specific mood, many interviewees indicated that they used Spotify's "radio" regularly. But the limitations of this comparison between algorithms and radio were constantly reached. For Rubén, Spotify's algo-rithmic radio metaphor quickly fell short. As he put it:

> Radio has something that these applications don't have: social interaction. I actually need the real live interaction and entertainment. Spotify, YouTube and all those [apps] let me adapt things to my liking, but [they] became so indi-vidualistic that you get bored.

Rubén's comment is reminiscent of integration dynamics explored in chap-ter 3. In his view, what Spotify lacked (that radio possessed) was an intrinsic way to combine algorithmic recommendations and the opinions of other people. For this reason, he found the radio metaphor wanting.

Playlists created by the company (rather than by users) are another exam-ple that reveals the centrality of affect in people's relationship with algo-rithms. Recent studies show that Spotify has shifted to promoting its own playlists in place of third-party lists since at least 2012 (Prey, Del Valle, and Zwerwer 2022). This shift points to the importance that playlists have come to occupy in Spotify's economic model (Eriksson et al. 2019).

The key for assessing these algorithmic recommendations was purely affective: it was based on whether users interpreted the moods and emotions these playlists conveyed as organic. For many interviewees, Spotify's curated playlists signaled the success of its algorithmic model of recommendation. Gabriela, the sound engineer, shared this view: "[Spotify] definitely knows what I like. They already know what the formula is to make a playlist for *me*" (emphasis added). For Gabriela, algorithms captured the moods and emotions that she typically sought to explore in her rituals. She attributed this outcome to a right interpretation of the input she had given to algorithms through her activities over time. By this logic, algorithms validated the time and efforts she had devoted to using the platform as the center of her affective life.

But other users seemed wary of Spotify and its algorithmic recommendations. Adrián, the public relations specialist, claimed that he never listened to Spotify's own playlists. When pressed to explain why, he noted: "What I feel is 'On the Bus Going Home' [the name of a playlist recommended to him] is not what Spotify says it is." He thus criticized the platform for producing affect that he perceived as artificial rather than organic. The evidence Adrián put forth to arrive at this conclusion is to have ritually experienced what being "On the Bus Going Home" actually meant.

The analysis of playlist creation as a ritual allows an understanding of how the cultivation of moods and emotions is legitimized as an exigence in people's life. This consolidates the myth of the platformed center in that it naturalizes the role of Spotify as an obligatory intermediary in the establishment of a utilitarian relationship between users and music.

BOREDOM AND PERSONALIZATION ON TIKTOK

The analysis of how users relate to Netflix and Spotify revealed how rituals are central in acting out the centrality of platforms in daily life. The case of TikTok allows the understanding of an additional dimension of the myth of the platformed center's operation, namely, how algorithmic personalization is positioned as a solution to the issue of boredom in daily life. Building on a growing number of studies, I begin by discussing boredom as a moral emotion. I then examine the rituals that TikTok users engage in to bring about algorithmic personalization and thus stave off boredom.

BOREDOM AS A MORAL EMOTION

Defining boredom, as Toohey (2019, 5) writes, is fraught with "confusion," "challenges," and "vagueness." Broadly speaking, scholars have envisioned boredom as "a sort of recognition of the absurdity of existence" (Ros Velasco 2017, 182) that results from the growth of leisure and wealth in capitalist, industrialized societies (García Quesada 2020; Hand 2016; Parreño Roldán 2013). Most researchers distinguish between boredom, tediousness, and laziness or sloth. Retana (2011), for example, defines tediousness as a generalized and overwhelming existential state, and sloth as a reluctance or indisposition to perform a specific task (usually associated with a form of work). In contrast, boredom "tends to have a definite object [. . . .] We are bored by something or in the absence of something" (Retana 2011, 182). Boredom is thus about specific, present situations (Toohey 2019).

Retana (2011) goes on to suggest that, as an emotion, boredom is triggered by contextual factors, leads to certain patterns of behavior, is characterized by specific bodily changes that are experienced as conscious feelings, and implies a certain degree of affective commitment. Moreover, Retana (2011, 186) contends that boredom must be theorized as a moral emotion to the extent that it "makes us aware of a certain dissatisfaction with our environment." This dissatisfaction typically arises from situations in which people are asked to comply with roles that don't satisfy them in relations of domination. Seen in this way, both individual and collective boredom "indicate to us the inability of our culture to produce meaning" (Retana 2011, 186).

Boredom was an important trigger for using TikTok among the people I studied. Yamila, a biotech engineer, captured a common sentiment that prevailed during the focus groups I conducted, when she stated: "I would use two words to define TikTok. First, 'boredom.' I use TikTok when I'm super, super bored. And the other word would be 'fun,' because I genuinely laugh a lot." In Yamila's account, TikTok linked together the expressions of, and solutions to, modern boredom.

For many interviewees, having to stay at home during the COVID-19 pandemic created "boring" conditions that led to using the app (cf Vallee 2020). In this sense, boredom is also a privilege or, as García Quesada (2020) puts it, a class issue. Mario, a 23-year-old international relations major, referred to the tendency of using TikTok during the pandemic as the "quarantine effect." He explained: "Many gave TikTok a 'redemption shot.' If they hadn't been so bored, many people would have deleted it. But since they wanted to consume content and didn't have anything else to do, they were like, 'I'll give it another shot'". A mix of pressure from close friends and the feeling that TikTok was an "overwhelming" cultural reference (in the words of one interviewee) in their social circles led many to install this particular app.

TikTok's relative novelty also fed a sense of originality. Nelly, who is 26 years old and works in health promotion, did not hesitate to define most social media platforms as "boring" when compared to TikTok. She said she downloaded TikTok at the start of the lockdown in Costa Rica, since she was dealing with insomnia, and other apps failed to offer her content that entertained her. Like Nelly, many users interpreted the pandemic as a context made of moments that needed to be filled with something new and different. In her case, this also included time in the middle of the night when she couldn't fall asleep.

This context intensified a common tendency among most interviewees: the practice of filling in time with the use of media technologies, most notably cellphone apps. The pandemic also made it necessary for many to look for a particular kind of content that could help them shift their attention away from the grim scenario. Said Mario: "TikTok is my oasis! It's a place of smiles in the midst of our disastrous and busy life. At least in my case, TikTok is a place of pure positive things." This statement aptly describes how individuals who had used TikTok for a longer time (even before the pandemic) felt about the app: it offered them a sense of escape from the boredom and monotony that characterized even other social media. Paulette, a 20-year-old college student, thus noted: "For me, TikTok is an escape, because I use it mostly when I'm fed up with my readings and my classes, and I don't want to go to Twitter or 'Insta.' I consume something where I don't even have to see comments!" Even as Paulette acknowledged the limitations of social media to resolve boring and tedious aspects of her daily life, she still maintained the idea that the solution lay in more technology use. For her, it was more about finding the right kind of content and apps than about changing the system that made the apps necessary.

Researchers have increasingly adopted the term "doomscrolling" to designate the practice of endlessly consuming content (including negative posts) on cellphone apps, to the detriment of people's mental health (Lupinacci 2021). In the case of news consumption during the pandemic, research shows that users strategically broke away from content to avoid the sense of being overwhelmed and emotionally drained (Ytre-Arne and Moe 2021a). Instead, my participants' strategy was to ritualize entertainment content consumption on TikTok to maximize the pleasure derived from "escaping" boredom through "stress-scrolling."

PERSONALIZATION AS RITUAL

As an emotion that needs to be resolved, boredom typically leads to specific actions. It "generates [a] mental space in real time that promotes mechanisms of transcendence and transgression [. . .] [it] creates a dearth of meaning that encourages the subject to fixate on the new" (Parreño Roldán 2013, 2, 8). To stave off boredom, many people turn to technology. As a 2015 study noted that, in the United States, 93 percent of smartphone owners aged 18 to 29 years old said they used their phone precisely to "avoid being bored" (A. Smith et al. 2015, 9). Authors have seen boredom as both a cause and a result of technology use (Gardiner 2014; Svendsen 2005; Turkle 2015). Referring specifically to the case of social media, Hand (2016) defined "digital boredom" as the continual filling in of time with activities that only exacerbate the feeling of being bored. It is the habitual rejection and simultaneous reproduction of boredom in order to gain a sense of stability "amid a world of conflicting time-scales arising from the human, the digital and the natural worlds jostling and interpenetrating one another" (Mosurinjohn 2016, 152). Some scholars thus argue that turning to technology in moments of boredom prevents modes of self-reflection (Turkle 2015).

Alternatively, Hand (2016) and Mosurinjohn (2016) propose a situated, practice-based approach that investigates rather than assumes how people relate to technology when dealing with boredom. Building on this approach, I consider these practices as rituals. Santiago, a business administration major, summarized several dimensions involved in the ritual of using TikTok:

> I always use it at night. Some days I don't use it but when I do, I spend too much time on the platform. I work from 6 a.m. to 3 p.m. Sometimes I use it at noon, during lunchtime, but I limit it because I know I can stay there for a long time.

Santiago thus used TikTok by carrying out certain actions repeatedly, in certain circumstances, and in a certain order.

For Santiago and other interviewees, the ritual use of TikTok took place at specific times of the day. Many indicated the use TikTok at the very end of the day. That was Georgina's case, a 22-year-old psychology student who said she always used TikTok at around 11 p.m., before falling asleep.

Using TikTok at the end of the day created a sense of expectation about the pleasure and entertainment that awaited users like Georgina in the immediate future. It also symbolized a form of reward for having endured the boredom, tediousness, and uncertainties of the day. Thus, Catalina, a public relations specialist, asserted: "I use [TikTok] when I no longer have any responsibility. I then give myself the right to say: 'OK Catalina, you can now check TikTok!'"

In these cases, many users preferred the bedroom as the ideal place to use the app. Santiago's comment, cited above, evidenced a common practice among users. Many said they did not use the app daily but rather waited a few days between sessions. Naomi, who is 21 years old and is about to graduate in communication studies, summarized it in the following manner:

> I don't use it on a daily basis, but when I do, it is a rabbit hole. I usually go to the "For You" page and content posted by the people I follow, so it's content that I know I'm going to like, and that's why I spend a lot of time using it.

Mousurijohn's (2016, 152) analysis of texting as a ritual offers a compelling explanation for this practice: "Disappearance is the precondition for reappearance. We must experience the loss of an object in order to refresh our desire for it." In a similar manner, when TikTok users were ready to refresh this desire for the app after a few days, they expected to use it intensively for extended periods. Although some users noted that they watched videos on their cell phones with other members of their household, using TikTok was usually done in an individual manner. But this practice was different from Netflix's individual rituals, discussed previously, which were carried out as a companion to a series of mundane activities. For Ernesto, a 22-year-old industrial engineer, using TikTok was something that he couldn't perform while doing anything else. He said he had to postpone it for a few days until he could devote himself completely to it for hours.

TikTok's appeal was explained as almost irresistible (hence the notion "rabbit hole" and using the app for hours "without realizing it"). As another interviewee put it, it was "like a drug." Sonia, who is a 22-year-old college student, offered a vivid description of this phenomenon:

> I used to use it for ten minutes when I was going to bed, mostly out of boredom. But then TikTok's algorithm began to "grab" [understand] what I like. And I started to see more and more things that I liked when the app began to know me.

Chapter 2 showed how users envision content personalization as an achievement, the product of an evolving relationship with algorithms. In addition, Sonia's observation reveals how users expected to achieve this result through ritual practice. Rodrigo, a library scientist, similarly stated: "I would define [TikTok] as a loop. Once you enter it, you cannot leave it. I am not complaining that I can't get out when I want to. It's so attractive that I don't mind being inside." The notion of "loop" implies the idea of repetition, something that has to be done constantly to achieve a result (in this case, an app that "can't" be put down). The myth of the platformed center operates here by promoting the need for staving off boredom without providing a definitive solution. Instead, TikTok promises a momentary solution to boredom that requires the repetitive enactment of practices to reach personalization. In short, the ritual use of TikTok needs to be constantly renewed to maintain personalization.

These rituals typically materialized in a set of specific routines when using TikTok. Users began their ritual appropriation of the app by always employing the same features, which allowed for securing content personalization. Rodrigo's case allows to appreciate these routines in operation:

> I watch [first] a few videos on the "For You" page. I then go to the "Inbox," watch a few videos there, and then return to "For You." It has never happened to me that I get bored. I never say, "*Mae* [man], I'm not seeing anything interesting anymore, I'm going to leave now." When I leave [the app] it is because I think it's been too much already. I say to myself: "You need to stop now."

The ritual use of the platform is then materially embodied in the act of swiping content, the gesture through which users keep content personalization going.

As noted previously, scholars have argued that digital media users tend to deal with boredom by reproducing the conditions that led to it in the first place (Hand 2016). Mosurinjohn (2016) contends that boredom is an inevitable product of the repetitive nature of rituals. Yet, as Rodrigo's words illustrate, the users that I interviewed were far from reaching that conclusion. In contrast, they envisioned TikTok as one of the few platforms that practically never bored them at all. To prevent boredom, there were always rituals.

CONCLUDING REMARKS

This chapter has traced the operation of what I have termed the myth of the platformed center, an adaptation of Couldry's (2003; 2012) concept of the "mediated center." Considering the cases of Netflix, Spotify, and Tik-Tok, I have argued that this myth operates through two main processes. First, it establishes the idea that daily practices, moods, and emotions require users' intervention. To frame it in Couldry's vocabulary, this category normalizes the notion that the world is made of concrete units (such as practices, moods, and emotions) and that intervening in their organization is a necessity.

Comparing these three platforms allows us to understand how technology companies partially differ in providing logics to justify this myth (Boltanski and Thévenot 2006). On Netflix, the category "Because You Watched" suggests the platform merely *extends* practices that were performed by users in the first place. In this way, Netflix positions itself as a neutral entity that mirrors previous user actions and behaviors. Spotify emphasizes instead the need to *cultivate* specific moods, that is, producing, capturing, or exploring them through ritual work and particular technologies (such as the playlist). And users turn to TikTok to stave off boredom through recommendation algorithms that will help them *escape* the burdens of daily life. In this way, personalization is normalized as an achievable goal. This category then translates into a set of practices that become meaningful for users through specific patterned actions (that is, rituals).

Second, this myth positions the notion that algorithmic platforms are the obligatory intermediary to carry out this intervention in daily life. This is achieved by normalizing the role of algorithms as solutions to the "problems" entailed by living a "media life" (Deuze 2011): deciding what movies or shows to watch next, how intensely to feel an emotion, or how to be entertained in the conditions generated by the monotony of capitalist societies or even during a pandemic. Couldry's (2009, 64) words can thus be aptly adapted to the case of algorithmic platforms: "A striking feature of contemporary [platforms] and [. . .] rituals is precisely the way in which they make natural (against all the odds) the idea that society is centred and the related idea that some [platform]-related categories are of overriding importance." In this way, platforms become the means to certify or legitimate the reality of certain practices, moods, and emotions. However, this does not mean that the myth of the platformed center goes uncontested. This chapter also accounted for instances where users challenged the operation of this myth. (This issue receives more attention in chapter 6.)

This chapter showed how users typically evaluate algorithmic recommendations against the backdrop created by the myth of the platformed center, as algorithms become infused with specific practical and affective meaning. In this sense, the myth of the platformed center must be understood as context dependent. In Costa Rica, performing rituals like consuming media content (such as movies, music, or videos) took shape against the background of a cultural mandate to spend time with people as an opportunity for bonding. To respond to this mandate, users turned to content mostly produced abroad. In neoliberal democracies, such as Costa Rica, ritualization also occurs in the conditions created by the legitimacy of the market ideal.

Bringing attention to this myth also helps expose some of the biases that have shaped the contemporary study of algorithmic platforms. The notion of "platformization" is a case in point. Scholars have used this notion increasingly to assess the implications of digital media in a variety of fields. In short, "platformed" is the new "networked"; it has gradually replaced the latter concept as a favorite umbrella term in the study of digital media. There is now talk of "platformed work and labour," "platformed discourse," "platformed interactions," "platformed racism," and "platformed cultural production" (Farkas, Schou, and Neumayer 2018; Hearn and Banet-Weiser 2020; Marton and Ekbia 2019; Matamoros-Fernández 2017; Nelimarkka et al. 2020). Despite the many contributions of this approach, an emphasis on the logic of platformization runs the risk of normalizing the myth that scholarship ought to problematize in the first place. The term "platformization" intrinsically assumes the claim for the legitimacy of platforms' power in centering social life rather than investigating the "actual complexity and uncertainty of the terrain in which [this power] operate[s]" (Couldry 2015, 621). As an alternative, this chapter has emphasized the centrality of rituals in broadening our understanding of the power of algorithmic platforms. In essence, I have argued that this form of power relies on the unabating reproduction of certain categories through rituals. Put differently, ritualized practices work to constantly enact the centrality of algorithmic platforms in daily life.

At its essence, the myth of the platformed center is about the notion that digital media are indispensable for bringing people together. Chapter 5 examines how users enact this notion in practice as they work with or against algorithms when recommending content to other people.

CONVERSION

As with many other users whose experience was discussed in chapter 4, Georgina, a 22-year-old psychology student at Costa Rica's largest public university, started using TikTok in early 2020, when social distancing measures were enforced in the country because of the COVID-19 pandemic. TikTok first caught Georgina's attention when she saw her sister perform various dances that she had learned through the app. And having to stay at home most of the time during the pandemic soon became, as Georgina put it, "quite boring." She thus decided to sign up for the app that so many of her friends were talking about. One year later, during a focus group, she seemed pleased with her decision. "I ended up loving it!" recounted Georgina, explaining that she found TikTok to be "always entertaining."

But Georgina's love for TikTok was not immediate. She did not find it obvious how to encounter "always entertaining" content. As she experimented with the app in those early days, her first reaction was to try to find videos created by people in Costa Rica. Her assumption was that this would help TikTok's algorithms find something she could relate to. Accordingly, she decided to search for content by using the words "Costa Rica." The results of this search surprised her. In her words: "I couldn't even describe it. I got things so, so strange!" Rather than a broad spectrum of possibilities from which to choose, as she expected to find, the app's algorithms showed her very narrow results: only specific kinds of people, living in the same parts of the country, always doing the same things. It became difficult to challenge what seemed to her to be a clearly biased pattern on TikTok's recommendations associated with the country. (Chapter 6 delves deeper on the significance of users' resistance to biased algorithmic patterns in recommendations.)

Georgina quickly learned three important lessons from this experience. First, that TikTok's algorithms would need some kind of guide from her to go beyond certain patterns. Second, that she would have to rely on recommendations from friends as well to receive the content she desired. And, third, that sending videos to others would actually improve their own experience with the app. Georgina explained: "[I started asking] people to send me videos to see if, by watching those videos and 'liking' the ones they had sent me, my algorithm would change and stop showing me so many [similar] things." Since then, Georgina has incorporated the habit of exchanging TikTok videos with others as a fundamental way of using the app.

Georgina's newly acquired habit can be considered an instance of *conversion*, or the process of connecting with a public through the display, sharing, and discussion of technology or its contents. Conversion dynamics occupied a central place in the original framework of domestication theory. Silverstone (1994) referred to conversion as a way for people to transform the private consumption of media technologies into a public issue. In Silverstone's (2006, 234) words:

> Conversion involves reconnection [. . . .] It involves display, the development of skills, competences, literacies. It involves discourse and discussion, the sharing of the pride of ownership, as well as its frustration. It involves resistance and refusal and transformation at the point where cultural expectations and social resources meet the challenges of technology, system and content.

In this view, conversion is a means to negotiate belonging, to claim status, to blur the boundaries that define consumption dynamics. The process of conversion turns media technologies into both the cause and the product of conversations with others (Silverstone 1994).

In this chapter, I argue for broadening this approach by further unpacking three notions involved in conversion processes: the self, the public, and the technology. Put differently, I define conversion as a set of intersecting enactments of the self, the public, and algorithms. Defined in this way, conversion transforms from singular to a plurality of processes. I situate people's relationship with algorithms within what I have called elsewhere "registers of publicness" (Siles 2017). By this, I mean "the articulated combination of mutually defining conceptions of self, publicness, and technology that characterize certain contexts and moments. They are deeply situated in that they both crystallize and help shape cultural formations and values" (Siles 2017, 13). This concept is meant to capture "the symbolic and material conditions in which performing the self in specific ways, through certain practices and the use of media technologies, becomes natural" (Siles 2017, 13).

Situating conversion within these registers of publicness means considering both the identity and relevance of algorithms as a partial result of how people conceive of their sense of self, their publics, and the role played by technology in connecting the two. Conversion thus articulates the self to others through algorithms, and it connects the self to algorithms through others. This approach provides a valuable addition to scholarly literature that has envisioned algorithmic appropriation as an individual act. In this alternative, I show that, even if users appropriate algorithms individually, they variously imagine themselves as part of wider publics and thus enact algorithms differently.

In what follows, I discuss three forms of conversion that are tied to different registers of publicness. In the case of Netflix, users performed a romantic self that constantly required expressing taste and knowledge through recommendations to others. These recommendations differed from those offered by algorithms in that they stemmed from a personal understanding of other people's lives and experiences. Users variously enacted technologies to materialize these recommendations to different kinds of publics: close ties, groups of friends, and "the people" in general. Next, I draw on Lauren Berlant's notion of "intimate public" to conceptualize conversion dynamics on Spotify. When users shared with others the playlists they had created, they performed an introspective and productive self that also knows how to have "a good time" when surrounded by others. The notion of algorithms as enablers of affective discovery was key in sustaining this register of publicness. Finally, I look at conversion in the case of TikTok. Unlike Netflix and Spotify, participating in public exchanges of videos was more explicitly framed as a reciprocal activity that required an attentive and responsible self. In this case, publicness referred to particular ties in users' social networks of "close friends." Conversion then became a collective technique through which the notion of "close friendship" emerged. Accordingly, users enacted algorithms as the guarantors of the successful operation of this process.

CONVERSION AS SHARING RECOMMENDATIONS ON NETFLIX

In Netflix's case, conversion can be interpreted through the lens of "sharing" as a constitutive activity of the social media imaginary (Belk 2010; John 2017; J. Kennedy 2016). John (2017) explains how the notion of "sharing" has become linked to the idea of making an internal state of the self available to others. This practice, John argues, became the ideal of interpersonal relationships throughout the first half of the twentieth century. According to John (2017), sharing practices in the case of social media inherited the premises of therapeutic discourse that began to be established by the 1920s. J. Kennedy (2016) refers to this dimension of sharing as "social intensification." She explains: "sharing is defined in relation to disclosure and affect, meaning to make oneself available to others through some form of sentiment articulation" (J. Kennedy 2016, 468). In what follows, I discuss the mutually defining notions of self, public, and technology that underlie conversion dynamics in Netflix's case.

SELF

Most scholars who have followed Silverstone's approach to conversion have emphasized issues of identity (Matassi, Boczkowski, and Mitchelstein 2019; Miao and Chan 2021). Haddon, for example, defined conversion as the mobilization of media technologies "as part of [people's] identities and how they present [. . .] themselves to others, for example, in how they talk about and display these technologies" (Haddon 2017, 20). In this view, conversion is primarily about oneself and about one's relationship with the public.

Identity issues were also key in the conversion dynamics of the Netflix users I interviewed in Costa Rica. To understand these dynamics, it is necessary to examine the underlying conceptions of selfhood that shaped users' recommendation practices. Streeter (2010) has shown the centrality of romantic forms of selfhood in the development of computing culture and the Internet in the United States. This liberal notion of the self—which Streeter calls "romantic individualism"—emphasizes ideas of stability, creativity, and singularity in self-performance (Siles 2017). Underlying this view is the "inside–outside" opposition that is constitutive of modern selfhood. Taylor (1989, 111) explains it with precision: "We think of our thoughts, ideas or feelings as being 'within' us, while the objects in the world which these mental states bear on are 'without'."

In a similar manner, Netflix users in Costa Rica performed a self that was primarily defined by its inner capacities (and obligations) to constantly create an original opinion about external cultural products. They envisioned the self as a creative force that expressed itself through original thoughts to prove its worth and uniqueness. "I just love to recommend," said Inés, a 25-year-old miscellaneous worker. She elaborated: "My sister and I don't have the same taste, but I still recommend things to her. I know she is not going to watch it, but I still recommend things to her. I just love to recommend!" For Inés, recommending was more about herself than about others. Recommending had a more profound effect on her and her self-understanding than on the person to whom she offered her suggestions.

How Netflix users conceived of themselves was perhaps nowhere clearer than in their explanations of *when* they felt the need to recommend content to others. These occasions reveal as much about how they thought about themselves as about how they valued the content they recommended. Suggestions were grounded on ordinary experiences with content watched on Netflix. Users had to watch and appreciate something before recommending it. To pass the "worthy of recommendation test," users offered various measures and indicators. Some spoke of "quality" as the main criterion for recommending content to others. Said Manuel, a 36-year-old computer scientist who worked in a public university in Costa Rica:

> There is so much content available [that] it can be difficult to find things of *good quality*. Thus, when I find something that I consider valuable, I share it. If [my wife and I] find something and it seems good to us, it [immediately] goes to Facebook. [Emphasis added]

In this example, the recommendation became the proof of the joint capacity of a couple to identify "quality" content. In a way, by setting this kind of standard, it is the user who is being tested rather than the content. When the content is found to be "of good quality," sharing it becomes an indication of the person's taste (or, in this case, a couple's taste). Thus, according to Manuel, "bad" content could not be recommended, because it would reflect poorly on them. It was their reputation that was at stake when recommending content to others.

Manuel also situated his recommendations in the context of content abundance (Boczkowski 2021). This suggests that, by being able to resolve this problem and recommend "quality" content to others, he was a knowledgeable person. This notion created fertile grounds for learning about algorithmic recommendations. Like Manuel, many users said they paid attention to algorithmic recommendations because they provided a sort of capital for conversion dynamics. In other words, algorithms offered them possibilities they could later transform into their own recommendations to others should the content pass their own tests.

Some users emphasized affective issues as their preferred type of trial: content needed to "touch" them and emotionally affect them to become worthy of recommendations. That was Luna's case (28 years old), who said: "If I think they [series, films] gave me pleasure when I watched them, then maybe others will also feel this." This type of proof required not only quality content but also a sensitive self that could both detect the feelings and emotions derived from watching a film or a series and also anticipate other people's potential reactions to particular content on Netflix.

The importance of issues of self in conversion dynamics was also illustrated by users who felt some sense of pride in being the first ones to recommend certain content. Elisa (39 years old) thus made sure to clarify that she had recommended *La Casa de Papel* (*Money Heist*) to others "before it became a trend" to watch and talk about it. During the interview, she emphasized that she had been able to recommend it because she had watched the series "before anyone else." In this case, the fact that this series became a trend confirmed her taste and an ability to anticipate content that others would enjoy.

PUBLIC

Interviewees envisioned sharing content on Netflix as an inherently good practice because of its importance in establishing interpersonal relationships. For most users, performing specific notions of a tasteful and knowledgeable self necessarily meant sharing recommendations, going so far as expressing a sense of obligation to recommend series and films to others, rather than keeping their opinions to themselves.

The nature of conversion dynamics in Netflix's case varied based on the specific type of public that users wanted to reach. On certain occasions, they recommended specific content primarily to strengthen their relationship with close ties. That was María's case, a 20-year-old university student who was watching a particular show at the time of the interview. She explained: "I feel like I'm watching it because I love them [certain girl friends] so much that [I say to myself]: 'I'm going to watch it just to have something to talk about together'." The affective value of having something to talk about with certain people (in this case, a group of friends) motivated María to incorporate certain recommendations in her life. Conversion thus became fundamental to her in ritually enacting meaningful relationships (see chapter 4).

Silverstone (1994) tied conversion to the interrelated issues of status, belonging, and competence. These issues were exemplified by Adriana's case, a public relations specialist. She employed WhatsApp to discuss Netflix content with an online group of women. Members of this group were located in Spanish-speaking countries and had never met in person, but they were linked by their common interest in writing. Adriana explained the dynamics of conversion involved in making this group work:

> I have a group on WhatsApp. We are approximately 10 [women] from different countries: Costa Rica, Mexico, Argentina, Chile, Spain, Bolivia, Ecuador, Venezuela, and Colombia. And Netflix is the same for everyone, except for the Spaniard. We all share many contents. One of us will say: "I watched this show and thought you would like it." If I watched something on Netflix and liked it, I would bring the subject to the group to discuss it.

In this example, Netflix was an object of shared attention across Latin America and Spain. The assumption of having relatively similar catalogs enabled conversations based on shared cultural references. This assumption is not minor: conversion required that participants in this group were convinced that others had access to the same content. For several interviewees, that was one of the things they liked the most about Netflix: it gave them the sense of commonality, despite the geographic distances involved. For Adriana, sharing recommendations was a way to signal membership in a social group that was relevant to her for both personal and professional reasons. Knowing specific types of content on Netflix was also a source of status: she felt she could establish a reputation based on the quality of recommendations she made.

In addition to close ties and strategic groups, interviewees shared recommendations with a more abstract or anonymous kind of public. Using a common Costa Rican (and Latin American) expression, they referred to this as *la gente* ("the people"). In these cases, sharing expressed a desire to transform aspects of the self (taste, knowledge of the platform) into an opportunity to influence others in their decision processes. Accordingly, users recommended content without having a specific person in mind. They shared their recommendations with whomever was willing to engage with them on social media. Elena, who works for a nongovernmental organization, thus defined herself as "*Black Mirror*'s preacher." When asked to explain why she felt so inclined to recommend this series, she noted: "I feel it is transcendental and important that everyone watches it." Elena's choice of words is telling: like a preacher, she felt the obligation to talk about a "transcendental" issue to an assembly of people whose faces she couldn't entirely recognize.

TECHNOLOGY

Sharing Netflix recommendations also relied on a fundamental view of technology, most notably algorithms. Users thought their recommendations differed from those offered by algorithms in fundamental ways. They argued that, although the final result was the same (that is, a particular cultural product made available to a person), the sources and rationale that justified their endorsements were entirely different from algorithmic recommendations. Romantic individualism was crucial to understanding this difference. For interviewees, one of the core differences was a capacity to know others on a "personal" level, that is, to be aware of their unique situations and experiences. This attribute then played what Foucault (1997a) would have called an "ethopoetic function": the romantic self was capable of transforming this knowledge of another person's life into a *personal* recommendation for them, one that reflected their *ethos* and spoke to their experiences in a direct manner.

This way of understanding the links between the self and technology has numerous precedents in the history of digital cultures. As noted previously, Streeter (2010) tied romantic individualism to the early development of the Internet in the United States. Siles (2017) also showed how this liberal notion of the self was key in the proliferation of blogging. In the mid- to late 1990s, it informed the practices of users (for the most part, although not exclusively, white, professional young men) related to the technology development field who created websites devoted to sorting the rising amount of content found on the Web. These users distinguished their websites (which they began calling "weblogs") from the operations of early search engines. To highlight this difference, they added relatively short comments that explained what made the links they recommended worth reading. These users envisioned such collections of recommended links as reflections of their personality, which could not be reproduced by what they thought were "*impersonal* lists of headlines" offered by search engines (Siles 2017, 47).

Similar premises informed my interviewees' approach to sharing Netflix recommendations with others. They were guided by the assumption that algorithms did not derive from first-hand knowledge of others or stem from a personal interpretation of their experiences in the context of a long history of exchanges. Consider how Ema, a 39-year-old college instructor, reflected on why she recommended content to others: "I would [share] if it were a very good movie, or documentary, or series that has something, a connotation that is important for another person's life." Ema invoked again the notion of self-approved content as the requisite for sharing with others. But the term "connotation" also suggests that recommendations needed to go beyond the most obvious meaning of a cultural product. Being capable of transcending the literal intention of a series or film by using it as a way to comment on a situation "in a person's life," as Ema put it, could only be achieved by someone who knew this person in detail. Ema illustrated the kind of situations that would warrant a recommendation: "If a friend is in a bad mood one day and there is a comedy or something I know about, I would say 'Hey, this is funny, maybe you'll laugh a little!'" On those occasions, recommending content was best accomplished by people who shared a history together and had the ability to identify those moments that warranted a specific suggestion. Marcia, a 19-year-old college student, summarized what most interviewees thought distinguished their recommendations from those made by algorithms: "It's because you *know* the person" (emphasis added). As will become clear throughout the chapter, this notion of the difference between a romantic self and technology pervaded people's interactions with other algorithmic platforms.

Through conversion, Netflix also gets embedded in a wider digital ecology. Users employ various technologies to discuss content with others. These technologies were strategically employed, depending on the kind of public the users wanted to reach. Whereas WhatsApp and messaging apps were preferred for sharing with specific individuals (close ties and groups of particular individuals), social media platforms such as Facebook and Twitter were used to reach *la gente*. This further complicates mutual domestication dynamics in that these discussions and recommendations are further filtered by the algorithms relied on by other technologies.

Netflix has consistently sought to use conversion dynamics on social media as a means to domesticate users. A first example was "Friends," an early initiative that had among its features the ability for users to view each other's lists and queues and make recommendations. In 2013, the company implemented "Netflix Social," which enabled users to employ Facebook to find out what people were watching on Netflix. These examples call attention to how much Netflix values "social data [as a] source of personalization" and its plans to use this information to algorithmically "process what connected friends ha[ve] watched or rated" (Amatriain and Basilico 2012b).

These initiatives have had relatively little success. In the media, Netflix has tended to blame users for resisting this form of domestication. In the words of the company's chief product officer, "we have had [. . .] major attempts at it and none of them have worked well. [. . .] It's unfortunate because I think there's a lot of value in supplementing the algorithmic suggestions with personal suggestions" (Hunt, cited in McAlone 2016). Similarly, *Business Insider* concluded that, "it's your fault Netflix doesn't have good social features" (McAlone 2016). But my interviewees seemed largely unconvinced that employing these features was of value in improving the recommendations they received. Thus, to this end, they turned to other technologies, which were much better integrated into their interpersonal relationship networks. Perhaps arriving at a similar conclusion, one of Netflix's most recent strategies has been to offer relatively generic features that allow users to share content from mobile apps.

In summary, conversion reveals how users understood themselves as social beings in relation to Netflix. It shows how they incorporated the platform and its algorithms into the networks of interpersonal relationships, where their identities, status, sense of belonging, and affect were partially defined.

CONVERSION AS PUBLIC INTIMACY ON SPOTIFY

As with Netflix, conversion was central to users' relationship with Spotify and its algorithms. Research has demonstrated that, in many ways, music is intrinsically social. For example, according to DeNora (2000, 109), music is a "device of collective ordering [. . .] a means of organizing potentially disparate individuals such that their actions may appear to be intersubjective, mutually oriented, coordinated, entrained and aligned." Thus, music has traditionally been an important means to enable feelings of community and belonging when consumed together or in public spaces.

In this section, I focus on conversion dynamics in the case of Spotify. Specifically, I examine how a group of Costa Rican users made public the playlists they created on Spotify (see chapter 4 for a discussion of the rituals involved in creating these playlists). Many users reported recommending music to others in ways that are similar to the dynamics discussed above for Netflix. Interviewees narrated several instances in which they felt compelled to "preach" on social media about the music and artists they love or send a specific song to someone, hoping it would speak to that person's life or situation, just like Netflix users did. But a focus on how they shared their playlists shines a spotlight on a different set of issues, namely, the centrality of affect and intimacy in their relationships with others and with algorithmic platforms. In short, I show that users make their playlists public as an attempt to create a shared affective experience.

To make sense of this process, I draw on Lauren Berlant's notion of "intimate public." Berlant's work showed the normative dimension of intimacy. According to Berlant (1998; Berlant and Warner 1998), having a "life" has increasingly translated into an obligation to have an *intimate* life. She articulated her definition of intimate publics as an alternative to Habermas' notion of the intimate sphere. For Berlant, public spheres are "affect worlds" in that emotions take precedence over rational thought, thus linking people together and, ultimately, shaping society (Jolly 2011). Unlike Habermas' view of "a sense of self which became a sense of citizenship only when it was abstracted and alienated in the nondomestic public sphere of liberal capitalist culture," writes Berlant, an intimate public "renders citizenship as a condition of social membership produced by personal acts and values" (Berlant 1997, 5). For Berlant (1998, 284), intimacy "emerge[s] from mobile processes of attachment"; it is fluid or "portable, unattached to a concrete space: a drive that creates spaces around it through practices."

Intimate publics are spaces where intimacy spreads. This occurs through the formation of a shared sense of identification, commonality, and belonging among strangers. Intimate publics emotionally connect people. In Berlant's (2008, 10) own words, "A public is intimate when it foregrounds affective and emotional attachment located in fantasies of the common, the everyday, and a sense of ordinariness." Intimate publics form when a fantasy of social belonging emerges through the shared experience of ordinary moods and emotions. When this happens, a "porous, affective scene of identification [flourishes] among strangers that promises a certain experience of belonging and provides a complex consolation, confirmation, discipline, and discussion" (Berlant 2008, 25).

Media technologies are key in commodifying intimacy and enabling the formation (and operation) of intimate publics. As part of capitalist consumer cultures, the media create markets of consumers under the premise that they can capture people's core desires through new content (Berlant 2008). Intimate publics also carry a normative dimension: they provide "the *proper* way of thinking about affect and emotion" (Berlant and Prosser 2011, 185; emphasis added). The formation of intimate publics is thus an achievement that serves a political and economic purpose. Yet Berlant acknowledges that intimate publics hold potential for political action. As Linke (2011, 16) summarizes it, Berlant "appreciates [intimate publics] as providing spaces for reflection, recognition, and a positioning of persons in their world, despite the ideologies that may be embedded in the culture of an intimate public." Thus, for Berlant, intimate publics can offer the possibility for nondominant groups to experience a sense of belonging. In one of the few studies that have used this notion for theorizing the case of digital media, Dobson, Robards, and Carah (2018, xx) positioned intimate publics on social media precisely within this tension. That is, they recognized the potential of intimate publics for providing an experience of belonging for nondominant people, but also recognized their role as "machines through which intimate practices are publicised, privatised, commodified, and exploited."

In short, intimate publics are "laboratories for imagining" (Berlant and Prosser 2011, 182). In what follows, I show how they worked precisely as sites for imagining the self, the public, and the place of algorithms in conversion processes on Spotify.

SELF

Conversion dynamics in the case of Spotify rested on specific notions of selfhood. As with Netflix, the assumption of Spotify users was that sharing was an ideal way to sustain interpersonal relationships by offering a window onto the self. Users normalized sharing as a practice with positive connotations. For many, recommending music to others felt like an obligation. Cecilia, a 30-year-old philologist, felt "selfish" (in her words) when she lacked the time to share music with others. Even those interviewees who said they preferred not to recommend music shared this assumption: suggesting music to others made them feel exposed.

To get a better sense of the particular notions of selfhood involved in conversion dynamics on Spotify, I conducted a content analysis of a sample of 100 playlists from all my interviewees. I selected various playlists from each participant in the study. (Thus, although not statistically representative, the sample captures the diversity that characterized the group of interviewees.) Through their titles and images, playlists evoked a sense of common, ordinary affect available to users that offers insights into how this group of Costa Rican users conceived themselves as well as how they conceived the nature of public intimacy.

Most playlists had names that evoked the performance of either introspective or productive activities. An introspective self emerged through the music selected for playlists such as "Rain" (a name that brought to mind a sense of contemplation while enjoying the sight and sound of tropical rain falling), "Straight from the Soul" (which pointed to the ethical substance involved in these introspection practices), and "Cassette" or "Vinyl" (which invoked the sense of nostalgia from the use and sound of previous media technologies). In this sense, playlist creation worked as a self-writing technique or what Foucault would have called a "technology of the self," a series of practices that allowed people to conduct a set of operations on themselves for the purpose of achieving certain goals (Foucault 1997b; Siles 2012b). Importantly, although created in a Spanish-speaking country, all these playlists were originally titled in English. This suggests that users sought to both perform a self with affinities in other parts of the world (namely, the United States) and to attract a public from these particular places. Accordingly, the overwhelming majority of music selected for these playlists came from artists with a worldwide presence rather than local bands. Whereas the music of English-speaking artists from the United States and the United Kingdom was spread throughout all the playlists, music from Costa Rican artists was typically isolated in specific playlists that often included a reference to the expression "*Ticos*" (short for Costa Ricans) somewhere in their titles.

Interviewees also created playlists devoted to fostering the formation of a productive self. The music selected for these playlists was meant to accompany professional activities in which this productive self could thrive. Typically, one user created a playlist named "Prog and nothing but prog" (also originally titled in English) to refer to the appropriate music that he selected for those moments when he programmed software (and nothing else). With this playlist, he sought to motivate himself to focus exclusively on professional tasks. Another person created a playlist named "Actividades" ("Activities").

He explained that the playlist was made specifically for listening while he worked in a particular organization.

Unlike playlists that were meant for practices carried out mostly individually, other playlists were explicitly devised for social activities. On these occasions, users employed playlist names with connotations of enjoyment and diversion. The premise in such cases was that the user needed to be introspective and productive while alone but had to know how to "have a good time" when surrounded by others. Such playlists included "Let's Dance" (originally titled in English and explicitly devised for entering into a "party mood") and "Weekends❤" (also titled in English by its creator to indicate that users needed to fully embrace the possibility to party on the opportunities afforded by the workweek).

PUBLIC

At times, users invited other people to create a playlist together. Rubén, a 39-year-old psychologist, thus recalled a playlist he created and maintained with a high-school friend with whom he has been exchanging music for 23 years. He explained:

> I told him: "I made the playlist and I'll share it with you, since we're music buddies ('compas' de música)." We named it "Oiga esto, mae" ("Listen to this, man"). The playlist was meant to showcase what the other was finding. It is fed by both. By sharing the playlist, it became social.

For Rubén, the ease of exploring and exchanging music together with his friend is what made Spotify, in his own words, "terribly addictive." On these occasions, what users shared was the task and pleasure of discovering music with someone else. For this reason, they turned to very specific individuals whose preferences in music they knew and trusted. Their ties to those people usually predated the formation of a playlist. More precisely, creating a playlist of music together was a way to sustain or extend relationships with a longer history. Social activities for which users prepared music that could allow them to sustain a particular mood is another example of co-creation. That was the case of Sofía, a 20-year-old college student who typically put together playlists for parties where those who planned to attend were able to participate in choosing the music and creating the playlist.

However, intimate publics formed primarily among strangers who shared—in the sense of having in common—the mood evoked by a playlist.

Sharing or making a playlist public meant opening an inner part of the self to others, as Eugenia, a 22-year-old journalism student put it. Eugenia's assertion is similar to those who shared Netflix recommendations. Sharing a playlist also meant opening an inner part of the self to others. Gabriela, a 21-year-old sound engineer, thus compared playlists to a form of personal "narrative." In her words:

> A narrative playlist is when I have a concept, a story or something that I want to tell someone, which I do through songs. So, I share them and say: "This playlist is . . ." [and I explain it]. All of them [playlists] are public, but I also share some with specific people, I share them "firsthand," so to speak.

Gabriela envisioned playlist creation as a form of self-expression. She distinguished between two kinds of "someones" to whom she addressed her song selections: a public made of strangers potentially interested in the narrative evoked by a playlist, and those familiar people to whom she sent the "narrative playlist" directly.

Additionally, a reflexive sense of belonging emerges through the display of the number of "followers" of a particular playlist. Spotify makes this number obvious for users, thus confirming the existence of an intimate public as soon as someone "follows" the playlist. Although many denied being interested in how successful their playlists were in attracting public attention, most people I interviewed seemed to be very aware of the number of followers they had gained. When Felipe, a 24-year-old computer scientist who works for a scientific laboratory, was explaining to me the playlist he made for playing one of his favorite video games, he interrupted himself to declare, with no small pride, "This is my playlist with more subscribers. It has 782." This kind of statement was relatively common during my interviews. The assumption was that, through the number of followers, they could quantify the affective effect produced by their playlists. In Felipe's view, 782 people could experience the *proper* emotional intensity associated with that video game.

Users worked hard to create these intimate publics. As Gabriela's words cited above suggest, some people shared their playlists on social media or messaging applications with comments that conveyed their emotional singularity. Carlota (a 30-year-old computer science student) said she typically shared playlists on Facebook to try to "inspire" others to listen to them. On other occasions, users turned to messaging apps to send a playlist's link to certain people, specifically, those "who can appreciate it," said Enrique, a 22-year-old public administration student. Berlant's (2008, viii) words are useful for making sense of these kinds of comments:

> What makes a public sphere intimate is an expectation that the consumers of its particular stuff already share a world view and emotional knowledge that they have derived from a broadly common historical experience. A certain circularity structures an intimate public, therefore: its consumer participants are perceived to be marked by a commonly lived experience.

Berlant aptly describes the premises and motivations of Spotify users who made their playlists available to others through the platform. In their view, the goal of sharing these lists was to allow others to experience the affective intensity enacted through playlists. This is similar to the motivation of the Netflix users discussed above, who stated that a particular content on the platform had to affect them emotionally to be worthy of recommendation. Carlos, a 20-year-old electrical engineering student, explained: "If [a playlist] made me feel good, I would like it to make others feel good in a certain way. I would like people to listen to new and different things and to explore the different sensations that [playlists] give." In this sense, playlists operated as a "sentimental intervention" that mobilized "a fantasy scene of collective desire, instruction, and identification that endures within the contingencies of the everyday" (Berlant 2008, 21).

TECHNOLOGY

Spotify enables a market of moods and emotions that can be consumed through the act of "following" playlists. Given the abundance of options, finding the proper playlist for cultivating moods is a work in itself. Users sought to find a representation of their affective interests in specific playlists. Nina, a 52-year-old audit specialist, explained how she typically dealt with the availability of public playlists:

> The app "boxes" things together; within the "drawers," there is a game of chance and probability that I might or might not like something. I start to see if there is something that catches my attention visually when I start navigating, something that sounds interesting to me. The title of playlists facilitates the trial and error.

Like Nina, other users employed the metaphor of "drawers" or "boxes" to describe how they understood Spotify's role in classifying playlists. This metaphor denotes a sense of unity and coherence in the musical substance of each playlist, but also a lack of openness. It also points to how users conceived of specific affordances in the process of conversion associated with Spotify. Algorithms played a central role in this process. In Netflix's case, people thought in general that humans could provide much better recommendations than algorithms could, because people know how to interpret specific situations and experiences in other people's lives; they only welcomed algorithmic recommendations when they felt these could allow them to build capital for conversion. Instead, for Spotify users, algorithms were enablers of affective discoveries that could help them to participate in intimate publics. Eugenia, the journalism student, noted:

> There is nothing more comfortable than going to Spotify and [to see playlists] already made to play them while you study, or to type: "I had a bad day, what should I listen to?" Since there is so much variety, it is easy to find something that fits your needs, so you don't have to go through that work [to create] and [can] just listen.

For Eugenia, the consolation for the ordinary experience of having a "bad day" came in the musical form of a ready-made playlist. These playlists also reminded users like her of the utilitarian role of music in performing a particular kind of self (one that devotes time and appropriate conditions for productive activities). By thinking of algorithms as enablers of affective discoveries, and in the name of convenience, users normalized Spotify's role as a producer of mood markets. Eugenia also illustrates the efforts of users who explicitly look for content outside of their "filter bubbles," a practice often ignored by the literature on algorithmic prediction (Pariser 2011). She explicitly relied on both people and algorithms not necessarily to find similar music to what she already listened to, but rather to explore new sounds that she didn't always know she "needed."

Intimate publics can also form through playlists created not only by users but also by Spotify itself. Some of these playlists have titles that reveal the commercial interest of intimate publics for the platform. There has been a special section devoted to "Genres & Moods" on the platform's browsing interface for several years. By the time I conducted the interviews with Spotify users, the "Moods" section recommended names such as *"Café, Libros"* ("Coffee, Books"), *"De Camino"* ("On the Road"), or *"FrienDeSemana,"* a play of words on weekend (*fin de semana* in Spanish) and "friend" (in English). Each playlist typically had a short description and a thumbnail. But regardless

of whether users "followed" other people's lists, or the ones created by Spotify, the playlist remained the privileged object of affective self-control in
their lives.

CONVERSION AS ATTACHMENT THROUGH TIKTOK

Whereas conversion materialized in the form of sharing dynamics on Netflix and public intimacy on Spotify, I argue that the notion of attachment
is most useful in examining the particular register of publicness that characterizes TikTok. For Hennion, attachment is "what links us, constrains
us, holds us, and what we love, what binds us, that of which we are a part"
(Hennion 2007, 109). Chapter 2 explored issues of attachment to TikTok
as a fundamental dynamic of personalization, which meant building a personal relationship with algorithmic platforms. In this section, I instead
examine mechanisms of attachment *through* TikTok. These are parallel processes: incorporating TikTok into interpersonal relationships solidifies the
attachment to the app and using TikTok more intensively becomes key to
sustaining certain relationships with others. Hennion's (2017b, 71) definition of attachment as "what we hold to and what holds us" points precisely
to this dual, mutually constitutive process. I begin exploring attachment
issues on TikTok by discussing how conversion is a key for getting people
interested in the app or, as Callon (1986) puts it, for "enrolling" people and
algorithms into an assemblage of video exchanges.

ENROLLMENT

It was common for people to become interested in or decide to install
TikTok because it seemed a convenient way to exchange content with
others. In such cases, conversion didn't follow other domestication dynamics, as Silverstone's (1994) early theorization implied, but rather preceded
and motivated them. Nicolás, a 22-year-old college student, explained his
experience:

> I didn't feel the need to download [the app] but there were some very good
> "TikToks" that my sister, my girlfriend, and my friends kept recommending.
> Since I didn't have the app, these recommendations came to me as videos on
> WhatsApp. It was because of this lack of comfort, of having to download many
> videos just to be able to watch them, that I downloaded the app.

As with Nicolás, the start of many users' relationship with the app was integrated within interpersonal relationships with a history. These cases are similar to Netflix users who emphasized the importance of knowing the other person's life as a foundation for conversion processes. Paulette's case is revealing in this respect. In her words, she started using TikTok because a friend recommended it insistently and promised to send her lots of videos if she would download the app. Paulette's friend anticipated how the app could work to extend their relationship. This anticipation functioned on the premise that this "friend" could know with precision what would be of interest to her and could incorporate communication through the app as part of their relationship. Installing the app became a means to automate and expedite these kinds of exchanges and thus enroll Paulette in the Tik-Tok assemblage. Rodrigo, a library scientist, equated his use of the app to a form of "chat" with others. He described this practice in the following manner:

> It has happened to me that, in some conversations, there are no text messages, there is only "TikToks" after "TikToks." That is the interaction. There are no reactions to the "TikToks" sent by the other person, and if there were texts, they would be ignored as we wouldn't see them buried among twenty other videos. So the app becomes a chat to send audiovisual [material].

The fact that Rodrigo and his friends did not feel the need to add any textual note stemmed from their conviction that exchanges on TikTok only added to broader conversations for which no additional context or precision was needed. Users often employed the notion of "accumulation" to refer to the abundance of videos they received. In this way, they expressed a notion of videos as a commodity that could be simultaneously possessed and exchanged. This interpretation was partly derived from the design of the app itself. Rodrigo elaborated: "The app's interface is *muy tuanis* (very cool). If you receive 10 'TikToks,' you play the first one and start scrolling the others as if it were a playlist from that contact [person]. That's why it makes sense to accumulate them [videos]."

"Tagging" also functioned as a technique for enrolling others. This practice reveals the mutual process of becoming attached *to* and *through* TikTok. This dynamic is nowhere clearer than in the incorporation of particular user practices that materialized ties with the app and with others in a single movement. Nelly, who is a 26-year-old worker in the field of health pro-motion, illustrated this process when she explained how she used TikTok:

> I always check first what my friends sent me and then I watch the videos that come out of [algorithmic] recommendations. I know that I will have messages from people who always send me things saying: "You have to watch these 'Tik-Toks'!" So, I know it's that time of the day when I'm going to laugh out loud because they are going to send me lots of stupid and interesting things.

The certainty of having "friends" who systematically sent her content that she would enjoy turned into a ritual in Nelly's day. (See chapter 4 for a discussion of rituals and their importance in enacting the power of algorithmic platforms.) In her case, the ritual started by checking messages from others rather than algo-rithmic recommendations, because they were sent by her "friends" and because these recommendations contained content she had come to like on TikTok.

SELF

In TikTok's case, conversion operates as an exchange of content with "friends" and a continuous enrollment through the platform. Participating in such kinds of exchanges requires performing a particular kind of self. Orlando, a dentist, described his experience with TikTok in a telling way:

> I send things that really make me laugh to about 15 friends who are on TikTok. The app shows them to me right there [in the app]. I feel that makes the experi-ence much more fun, because they respond to me, and the "chat" [referring to the app's Inbox] offers you the possibility to "like" the messages. So they get to laugh, I get to laugh, and we end up establishing a conversation through the app.

Conceived of as a "conversation," users felt they couldn't stay "silent" and were required to respond to others. A social interaction norm applied to the "chat" forced them to avoid being impolite and to respond to others when interpellated by them. In doing so, they expressed how they understood the self and the public involved in their TikTok video exchanges.

Orlando's comment reveals the similarities between the concepts of self-hood on Spotify and TikTok. As with the people whose practices I dis-cussed in the previous section, the TikTok users that I studied evinced a need to send videos that reflected how they could have "a good time" when surrounded by others. Explanations of what kind of content they shared frequently included references to instances of enjoyment, amusement, and relaxation. That was the case for users like Nelly, mentioned above, who emphasized that the videos they exchanged guaranteed they always "laughed out loud." Whereas user approval in Netflix's case centered on the notion of "quality" (and how it revealed the taste of the recommender), on TikTok it revolved around ideas of enjoyment and fun.

But unlike Netflix's and Spotify's cases, conversion dynamics on TikTok were understood much more explicitly as reciprocal exchanges. As Belk (2010) notes, although reciprocity is not necessarily expected when gift giving, it is customary in its practice. Users interpreted the app's design as an affordance that invited them to respond to each incoming video as part of a growing conversation. Mónica (20 years old) thus said that she could not bear the feeling of having unread messages on TikTok's "Inbox" feature, the space where videos sent by others are displayed. The sense of being a responsible self thus created pressures that materialized in the obligation of sending something back. The words of Catalina, a public relations specialist, also illustrate how sending videos was interpreted as an obligation of the self to the other: "I can't see something good and not send it to anyone. If I'm willing to give it a 'like,' if it's something really good, or something that makes me say 'Wow!', then it's something I have to share." Users incorporated the obligation of sharing to such a degree that, when failing to meet it, Camila, a political scientist, did not hesitate to refer to herself as "super selfish." As with Netflix and Spotify, the person must approve of the content before offering it to others.

Once the content has passed the user's own filters, then the person "has" to share it with others. Ernesto, an industrial engineer, summarized this practice and its significance in his interpersonal relationships:

> Since I receive many videos, I like sending them to others as well. As soon as I see a good video, I send it to others immediately. I generally start using TikTok by watching the videos I received from others, and then I proceed to do my end of the deal, so to speak. It's my turn to watch and send to others the good ones I saw. It becomes an endless list of videos that I receive and send. I know that others have watched them because, when we see each other elsewhere, we talk about them.

Ernesto's words almost suggest a sense of sacrifice in fulfilling this reciprocal obligation. He also highlights the centrality of conversion in his practices. He framed conversion as a two-way "deal" by describing his own use of the app as a sequence composed of two equally important parts: both receiving and sending. His use of the app wouldn't work if any part were compromised. Once this balance was reached, Ernesto said, using TikTok served to sustain quality relationships with others and extend these interactions both temporally and spatially.

PUBLIC

A specific concept of the public underlies the conversion practices discussed thus far. TikTok users did not send videos to anybody who might be willing to see them (like some did in the case of Netflix); nor did they expect that algorithms would find an appropriate audience for them (like many playlist creators did on Spotify). Instead, they sent these videos to very specific groups of people, often in an individualized manner. The logic of this conversion process was to singularize videos that had been circulating for everybody; it was about transforming algorithmic recommendations for *many* into a singular suggestion for *some*.

The public in TikTok's conversions can be characterized as the strongest ties in the users' social networks. It is against the backdrop of this notion of publicness that performances of an attentive and responsible self who reciprocates video exchanges is performed. During conversations with participants in focus groups, only one person (a 20-year-old college student named Pilar) acknowledged that she had created her own videos to "try to become famous" on TikTok. After uploading videos that she thought had potential for viralization but that failed to achieve this goal, she gave up on her original plan. Pilar clarified this change in her behavior:

> If I had gone viral, if I had gained more "followers," then I would have kept doing it [publishing public videos]. But now my "followers" are my friends. My sister was the one who told me, "Pilar, you're in my 'For You' [page]!" I almost don't make my videos public nowadays.

The notion of "follower," well established in social media lingo, further substantiates the view that the public TikTok is best understood as ties in a social network.

The overwhelming majority of people that I interviewed said they had created videos to send them only to specific individuals. These videos were meant for what users called their "closest friends." Mario, a 23-year-old international relations major, said that he shared videos he created with people beyond his closest friends only "occasionally." He explained:

I've never meant to publish them, to be honest. I only send them to very specific people when I think of something in particular. For example, my girl friend and I make comedy skits but only share them between ourselves.

Mario used the Spanish word *publicar* (to publish), which makes explicit the link between sending videos and his notion of publicness. For him, making a video public meant specifically reaching out to people beyond his network of social ties. In this view, sending them to "very specific people" allowed him to stay within the confines of the semi-public or the interface between the public and the private.

Although TikTok's design invites people to exchange videos as part of the app's own functionalities, users said they have also employed other platforms, most notably Instagram's "Close Friends" feature or WhatsApp, to be able to control the specific group of people who would have access to the videos they had created. That was the case for Catalina, who noted during a focus group: "The only time I have done that [making a video she had created public] in my entire life, I uploaded it to 'Close Friends' on Instagram so my girl friends could watch it." Seen in this way, conversion on TikTok is a collective technique or process through which a sense of "close friendship" emerges and is performed. It involves multiple actors (both people and technologies) linked together in particular ways. A certain notion of "friendship" is the product of how TikTok mobilizes an assemblage of attachments.

TECHNOLOGY

Algorithms play a specific part in this register of publicness. Unlike Netflix's algorithms, which people considered as an alternative to their recommendations, users indicated that they relied on TikTok's algorithms to sustain their attachment to others. In other words, users enacted algorithms as guarantors of their attachment to others through the app. The underlying premise behind this notion was that algorithms were capable of detecting patterns in the types of content shared with others in order to reproduce them. Catalina thus noted: "If I want the app to keep recommending some kind of content, I 'like' it and share it, so I can get similar things. It's a way of saying, "*Mae*, I want more of this!" This statement combines the sense of personalization examined in chapter 2, expressed by her way of addressing the platform as "*mae*" (the quintessential Costa Rican expression for "man" or "dude"), and conversion dynamics. Catalina thus envisioned conversion as a form of directly influencing algorithms in specific ways: I share, therefore I get. Many users that I interviewed envisioned conversion as indispensable in this process of personalizing the app's content.

For users, algorithms learned from both parts of the conversion "deal," that is, both when they sent content to others and when they received it. Isabel, a 21-year-old college student, explained why she thought that TikTok's "Inbox" (the space for exchanges with others) shaped the "For You" page (where algorithmic recommendations reign):

> I don't watch anime. But a friend of mine does watch it and keeps sending me anime and cosplay "TikToks," which I don't care about. She sends them and sometimes I get them on my "For You." Therefore, I think that whatever others send you can influence the algorithm that [regulates] what will be on your "For You."

According to users, not only could algorithms recognize the significance of conversion dynamics in the use of the app but they would also value these reciprocal exchanges of videos over other data sources employed to personalize content. Isabel, whom I just cited above, was thus convinced that her video exchanges would be "prioritized" (in her words) in her "For You" page.

In summary, if algorithms played their part as guarantors of the TikTok assemblage, then users trusted that their attachment to the app and to others was secure. Yamila, a biotech engineer, offered a telling illustration of this process. She imagined a hypothetical scenario that reflected common conversion experiences in her life:

> It is somewhat difficult to maintain a conversation, no matter how close of a friendship. If I haven't talked to a friend throughout the day, but she sends me a message on WhatsApp that says, "*Mae*, get on TikTok right now!", I would say, "Ok!" If I logged on, there would be 40 TikToks from her. I would ask her, "Which one do you want me to watch?" And she would say something like, "The one with a guitar." If I were to watch that one, we could start talking over there.

In Yamila's account, conversion on TikTok was like sustaining a long conversation that spread through both the private and the public and that bridged online and offline. As this conversation unfolded, her attachment to TikTok and to her friend would grow.

CONCLUDING REMARKS

This chapter positioned users' relationship with algorithms as a public issue. To this end, I argued for situating conversion dynamics within a larger analytical framework that built on mutually defining notions of self, publicness, and technology. This approach critically interrogated views of the self, the public, and algorithms that tend to be normalized in dominant accounts of conversion dynamics in the scholarly literature. Rather than considering the case of a fully constituted self that displays or uses technology to reach out to the "outside world," the approach espoused in this chapter turned self, public, and algorithms into products of one another (Siles 2017). Conversion is the name of this game; it is the logic that links these notions together in distinct ways. Examining how these connections emerge and are enacted in practice becomes the crucial task of inquiring into conversion issues.

The cross-platform comparative method also allowed an appreciation of the differences in how users enacted algorithms in their practices of conversion. In some cases, conversion was a process of expressing an opinion that revealed an inner feature of the self (such as taste or knowledgeability) under the cultural premise that sharing was indispensable in the formation of personal relationships. Users turned to the trope of romantic individualism and views of the self as an intrinsically expressive entity that could articulate unique opinions in ways that algorithms could not. This notion was particularly salient in Netflix's case. Interviewees combined multiple technologies to reach both groups of specific people as well as a wider network of people. They defined themselves through recommendations that sought to establish their identities, status, sense of belonging, and affect within networks of interpersonal relationships.

Conversion was also tied to issues of intimacy or, as Berlant (1998; 2008) contends, to the public exigence of living an intimate life. This demand was met with enthusiasm by Spotify users who shared with others the playlists they created, thus forming an intimate public. In this way, a scene of identification could arise among strangers that offered experiences of belonging and consolation to introspective, productive, and enjoyable selves. In this enactment, algorithms could enable the discovery of the appropriate public intimacy. The centrality of conversion dynamics on Spotify can thus be explained as an outcome of mutual domestication: they combined longstanding practices of sharing music with others and affordances built into the platform to make this possible in an "natural" way and thus extract data from users.

Finally, as TikTok's example revealed, conversion also meant attachment. In this case, conversion was enacted as a heterogenous technique of relations between people and algorithms, through which the notion of "close friendship" emerged and was constantly updated. For users, algorithms acted as guarantors that this process would operate successfully. Thus, the role of algorithms in people's lives varied as it was tied to (while also shaping) issues of self-performance and publicness.

In all the cases I examined, users held the conviction that a core capacity of the self was to know how to recommend content that could speak directly and uniquely to another person's life and situation. The groups of Costa Ricans that I spoke with thus demanded of themselves that they be sensitive both to their media environments (for possible content to recommend) and to other people's lives (for possible moments that required a recommendation). Accordingly, algorithmic recommendations were welcomed when users felt they could expand their capital for conversion, that is, could offer them options they could later recommend once self-approved. This notion that content (such as films, series, music or user-created videos) had to prove its worth before being recommended to others was also common to users of all platforms. People invoked different criteria to justify what made certain content pass the test: they mentioned "quality" when it came to performing certain identities (a knowledgeable self or a person of taste); affect when they sought to determine the content's potential for enabling feelings of identification, commonality, and belonging; or notions of amusement and entertainment when they wanted to inspire others about how to have "a good time."

By positing conversions in the plural, I have also sought to stress people's capacity to inhabit multiple publics through and with algorithms. In short, going back to issues developed in chapter 1, people enact multiple realities in practice. Applied to conversion processes, this means that users can simultaneously participate in multiple registers of publicness (including those expressed by the notions of sharing, public intimacy, and attachment) through a set of practices and relations with others (both people and algorithms). Thus, to paraphrase Law (2015), it is not that people live in one-public-world but rather that they live in multi-public-worlds or a set of intersecting publics (a pluriverse, as Ingold [2018] defines it).

The notion of politics is embodied in conversion issues. As Berlant and Warner (1998) contend, the notion of public intimacy implies a political potential. In a similar manner, claiming that realities (such as registers of publicness) are an enactment rather than "a destiny" invites a consideration of "ontological politics" (Law 2008, 637). Chapter 6 addresses this dimension more explicitly by focusing on issues of resistance.

RESISTANCE

Compared to power and control, resistance to algorithms and datafication has received significantly less attention (Brayne and Christin 2021; Fotopoulou 2019; Velkova and Kaun 2021; Ytre-Arne and Moe 2021b). In part, this difference expresses a broader lack of relative interest in resistance in the humanities and social sciences, already noted by García Canclini (2013, 4): "while conceptions of power and its movements have become more complex, notions of resistance exhibit astonishing inertia." This problem is particularly acute in studies on datafication: issues of algorithmic governance and control have been overemphasized at the expense of studies on resistance. This omission runs the risk of reifying a duality that limits our understanding of datafication. As Foucault has put it, this is because resistance is immanent to power. In other words, power and resistance are co-constitutive and dialectical (Mumby 2005). Thus there is a need to examine the complex dynamics through which (algorithmic) power and resistance intersect in datafication processes.

In the research that has been undertaken, scholars have examined two main ways in which users manifest or enact resistance to algorithms. First, they have focused on challenges to forms of labor associated with the so-called "gig economy" (Allen-Robertson 2017; Rosenblat 2018) or to algorithmic management in organizational settings (Brayne and Christin 2021; Kellogg, Valentine, and Christin 2020). These researchers have usually focused on the formal and informal practices and tactics by which individuals (often as workers) express their dissatisfaction with algorithms and attempt to alter their operation. As Velkova and Kaun (2021) note, the emphasis here is on "correction" (in the sense of exploiting existing shortcomings) rather than on producing an alternative system of datafication. Ferrari and Graham (2021) similarly speak of "fissures" in algorithmic power created through creative and playful practices that open up possibilities for agency.

Second, researchers have examined specific forms of data activism led by specific social groups or movements (Dencik, Hintz, and Cable 2016;

Ricaurte 2019; Sued et al. 2022). These studies have made a valuable contribution to understanding how resistance to datafication can lead to important transformations that favor human dignity and equality. For Couldry and Mejias (2021), achieving such a goal requires a resistance that is global (as opposed to strictly local or national), "double" (in the sense that it involves both practical actions and struggles at the level of knowledge and cognition), inclusive (and should therefore include a wide range of people and concerns), and focused on transcending the dominant values of Silicon Valley.

In this chapter, I examine the resistance that is present in ordinary life. Of course, this form of resistance has similarities to some of the work described earlier in this chapter (most notably the focus on practices that seek to express a discomfort with algorithms rather than attempting to change them.) However, there are some important differences between those forms of resistance and the practices of the people that I studied in Costa Rica. These differences lie in the context of resistance: what counts as resistance might differ from setting to setting. Organizational settings significantly change the context in which resistance arises. For example, the users that I analyzed did not perceive algorithms as their "boss," "manager," or any form of authority that could shape or control their incomes. Ordinary practices of resistance also differ from explicit attempts to change the pillars and logic of datafication through collective action in terms of their nature, public visibility, and forms of organization. Concurring again with García Canclini (2013, 7): "We understand little about historical changes in culture if we reduce them to the choice between resistance and domestication." Accordingly, in this chapter, I do not focus on how these users changed the operation of platforms such as Netflix, Spotify, or TikTok. I explore instead the cultural terrain of meaning-making in which both resistant practices and emotional responses arose that could have the potential to change datafication. I consider these practices and attitudes toward algorithms as forms of resistance in that they question the logics and premises on which datafication rests (Mumby 2005).

More precisely, I examine resistance to algorithmic platforms as a matter of "infrapolitics," or the substratum of more explicit forms of resistance (Marche 2012b). First advanced by Scott (1990), the notion of infrapolitics designates "a wide variety of low-profile forms of resistance that dare not speak in their own name" (Scott 1990, 19). For Scott, infrapolitics offered a way of overcoming the narrowness of framing political life as either consent or as open rebellion. He defined infrapolitics as "the prevailing genre of day-to-day politics [. . .] a diagonal politics, a careful and evasive politics [. . .] made up of thousands of small acts, potentially of enormous aggregate consequence" (Scott 2012, 113). Infrapolitics are the practices and gestures that express a form of resistance operating at the interstices of everyday life but may lack political articulation. In his analysis of Malay peasants, he illustrated this genre of politics with actions such as foot-dragging, feigned ignorance, sabotage of crops, and gossip (Scott 1990).

Because infrapolitics operates "beneath the threshold of the 'political'" (Marche 2012b, 5), some authors have challenged its relevance. Contu (2008) labeled this type of practice "decaf resistance," thus suggesting that, compared to some form of "real" or "full-blown" resistance, infrapolitical practices fail to change, challenge, or disrupt forms of domination. Yet a wealth of research has recently contributed to shining a light on the significance of infrapolitics (Baudry 2012; Courpasson 2017). These studies show how infrapolitics can potentially have "aggregate consequences," as Scott put it (2012, 113). In this sense, infrapolitics is "the subterranean magma of [opposition]" (Johnston 2005, 113). Infrapolitics can also be the ideal way to resist and can even be preferred over established forms of collective action (Marche 2012a). Regardless of whether it eventually leads to fully fledged political action, infrapolitics also matters in its own right: it represents a claim for people's agency, autonomy, and dignity (Marche 2012b; Mumby et al. 2017). As Brayne and Christin (2021) have shown, an account of how people relate to algorithms would be incomplete without understanding these practical strategies of resistance.

Building on these ideas, in what follows, I consider resistance practices that characterize the relationship between Costa Rican users and algorithmic platforms. I begin by discussing how Netflix users challenged what they perceived as Netflix's biases in algorithmic recommendation. To them, these biases were most obvious in the content that was both created for and recommended in Latin America. Netflix users claimed their identity against the backdrop of these biases. I examine next how Spotify users reacted to the platform's efforts to domesticate them into becoming paying customers (rather than "basic" users who employed the app for free) and the pressure to make algorithmic recommendations as the default of music consumption. Instead, they argued for having the freedom to establish their autonomy in relation to the music they chose and listened to. Finally, I analyze resistance efforts that were centered on highlighting TikTok's political project. I discuss users' insistence on the simultaneous need to delete content promoted by TikTok while promoting issues that were typically suppressed by the app's algorithms. By focusing on these instances of infrapolitics, I do not mean to romanticize resistance. Instead, my goal in this chapter is to offer more nuanced understandings of the relationship between power and resistance than those offered by the dominant accounts of datafication.

CLAIMING IDENTITY AGAINST NETFLIX'S BIASES

Although interviewees widely appreciated Netflix, they also criticized some of its features. This critique permeated users' relationship with recommendation algorithms and thus ended up shaping the mutual domestication process. Resistance to Netflix centered on two distinct but related issues: the kinds of content offered by the platform and the mechanisms it employed to promote this content (most notably algorithms). In this section, I discuss these two issues and how users expressed their resistance (that is, by ignoring the recommendation as a form of opposition to being treated as a consumer "profile" rather than a person).

CHALLENGING NETFLIX'S "EXAGGERATIONS": CATALOG AND ORIGINAL PRODUCTIONS

Resistance to issues of content focused primarily on what users perceived as Netflix's tendency to exaggerate, that is, to exacerbate certain trends in content production and distribution. Users found examples of this tendency in the "Original" series that Netflix created for all its users and in those specifically made for Latin American audiences. In essence, this form of resistance expressed a feeling of being treated as a passive consumer who could be easily duped into falling for stereotypical recipes of commercial success.

Many interviewees seemed to appreciate Netflix's own productions and considered them a new standard of quality in contemporary audiovisual culture (see chapter 3). But they also reacted against what they perceived as obvious biases in Netflix's approach to content production. This dissatisfaction was nowhere clearer than in users' assessment of Netflix's "Original" productions for Latin America. Ariana, who is 37 years old, offered a vivid illustration of how Netflix users both admired the company's technical merits and disliked the commercial goals of its productions:

> What I like the least is the content in Spanish. It is very focused on crime and *narco*. I believe that [Netflix] could exploit so many other Latin American stories. I watched *Narcos Mexico* and it was very good, I enjoyed it. But that doesn't mean that I want to keep watching everything related to drugs throughout Latin America.

As noted in chapter 1, since 2015, Netflix has produced original content in Latin American countries with large local production industries and vast consumer markets (primarily Brazil and Mexico, but also Argentina and Colombia). The platform has almost no content produced in smaller Latin American countries, such as Costa Rica (except for a few films). Most interviewees claimed to have watched (and appreciated) Netflix's Latin American "Original" productions. Most mentioned specific series they enjoyed (particularly *3%* and *La Casa de las Flores*). But they also felt the emphasis on large markets made it difficult for the company to transcend stereotypes associated with Latin America. In short, users generally agreed with Ariana, quoted above, in that Latin America had more stories to offer than the local history of drug trafficking in each country.

In addition to the overall selection of certain themes, users also reacted negatively to the aesthetic treatment that those themes received in Netflix's productions. Again, users perceived that Netflix exacerbated content cues that ended up turning series into a caricature of genres. In this way, in interviewees' own words, "romantic comedies" were turned into "cheesy" stories, Latin American series ended up acquiring a *telenovela* "flavor," and the story of Pablo Escobar and drug cartels became a romanticization of blood and violence. More broadly, users suggested that such trends evinced core biases in Netflix's strategy of original content production. The problem, according to interviewees, was that the company emphasized quantity over quality in its series; in this view, Netflix preferred to release numerous shows rather than securing quality in fewer ones. Users contrasted this approach with the "quality" they said they encountered on HBO or even Hulu.

A widespread criticism among interviewees was the lack of certain content in Netflix's catalog for Costa Rica, which was manifested in content that was *not* recommended to them. The desire for cultural proximity with the global north, discussed in previous chapters, was also a key concern in Costa Ricans' resistance to Netflix. My interviewees assumed that there was a singular catalog for Latin American countries. Many disputed the criteria employed to restrict content to countries of the region, including Costa Rica. Adriana, a public relations specialist and business administration student, expressed this discomfort with precision: "If there is something that truly bothers me, it's the fact that the Latin American catalog is completely different from the one in the United States. It bothers me that they have Latin Americans pegged as 'not class A' clients." A quality distinction was thus made between content offered in the United States and products available in Latin America. This dissatisfaction centered on a discrepancy that users deemed arbitrary: although they paid the same prices, they didn't have access to the same content available in the United States.

According to some estimations, Netflix's catalog in Costa Rica is almost 30 percent smaller than that in the United States (Moody 2018). But this complaint was not entirely based on quantitative criteria. Natalia, a college student, expressed this in a typical manner: "I went to the United States and [Netflix] had everything! All the series! The content for Costa Rica is tiny. When I log in on Netflix in Costa Rica, I always spend 30 minutes searching and keep finding that 'This title is not available'." Some people offered examples of trips to the United States where they noticed catalog differences. Others indicated that they often received recommendations from friends in the United States, but that they couldn't access the shows because they weren't available in Costa Rica.

Moreover, users didn't always consider Netflix's Latin American "Original" series as an alternative to content uniquely available in the United States. Marcia (19 years old) captured with precision the resistance to both Netflix's proneness to "exaggeration" and her preference for content from the United States: "I've watched perhaps three [Latin American "Original" productions]. And that's too many already! I wouldn't watch more, because they are way too exaggerated. The thing that appeals to me the most are American shows." As both Natalia's and Marcia's words suggest, what really bothered users was not having the same options that Netflix users in the United States had. Costa Rican users valued English-speaking content, because it allowed them to be a part of global conversations spurred by Netflix; it allowed them to be a part of the "Netflix phenomenon" as a global process (Lobato 2019). In this context, users interpreted catalog differences as a form of exclusion. Viviana, a 20-year-old college student, did not hesitate to use the word "segregation" to criticize this issue. This type of comment offers clues to better understand claims of cultural imperialism, because such remarks partially complicate the notion that the North–South division doesn't matter in datafication processes (Couldry and Mejias 2019).

BREAKING INTERPELLATION: BIASES IN NETFLIX'S
ALGORITHMIC RECOMMENDATIONS

In addition to Netflix's perceived exaggerations in creating, developing, and selecting content for the Latin American region, users often resisted the platforms' algorithmic approach to recommendations. This form of resistance can be considered to be a reaction to Netflix's algorithmic interpellation (see chapter 2). The spell of interpellation was usually broken for three interrelated reasons: the lack of interest in the recommendations offered to people; the difficulty in understanding why users had obtained certain recommendations; and the detection of biases in the promotion of certain content. Underlying these forms of interpellation resistance was again the realization that Netflix operated by considering users not as people but rather as profiles.

The most prominent reason to resist algorithms was the feeling that algorithms "failed" constantly, that is, that the content suggested to users was not relevant to them. On these occasions, interviewees considered algorithmic interpellation to be wrongly addressed. Fernanda, who works as communications director of a transnational enterprise, thus indicated: "[Netflix] comes and recommends something and I'm like, 'You have too much information about me to recommend things that you know I won't like. Then why am I giving you my information?'" Fernanda thus spoke back to the Netflix subject to complain about its failures to fulfill the expectations derived from interpellation. This attitude is reminiscent of Ytre-Arne and Moe's notion of algorithmic "irritation." They explain: "user perceptions of algorithms are not fully grasped by the notion of 'resignation.' [Irritation opens up] a realm for user agency in criticizing algorithms, actively noticing their imperfections, rather than accepting them as seamlessly integrated into media experiences" (Ytre-Arne and Moe 2021b, 820). Yet many users also blamed themselves for Netflix's "failures." In a typical manner, Rosa, a 30-year-old psychologist, said it was her fault if she was getting "wrong" recommendations, because she was not "teaching" the algorithm well enough. In this way, the obligation to provide Netflix with more explicit feedback was incorporated into specific user practices.

A second source of resistance to algorithms stemmed from the opacity of the platform. Users constantly tried to understand how the platform and its algorithms worked. Jorge, a 34-year-old master's degree student, explained his own theory: "What I think is that [Netflix] makes recommendations based on what I've already seen." However, the lack of clarity on how users received specific recommendations (and not others) generated some frustration. Jorge elaborated: "Suddenly, they 'throw' [suggest] things that make you wonder: 'How did I get this? Why do you suggest this?' Some recommendations don't fit with what I've watched [previously]." Jorge thus felt surprised by recommendation criteria that remained incomprehensible or incomplete. In other cases, resistance derived from what people interpreted as a misreading of the inputs given to Netflix through consumption practices. Adriana, the public relations specialist, maintained: "Sometimes it's completely wrong [*se la pela*] with the recommendations and I [ask myself]: 'How can Netflix not know me at this point?'" Adriana's reflections prompt two observations. First, users expected concrete results after investing time and effort in interacting with the platform. Second, although how Netflix operates is never fully understood, the expectations of the recommendation mechanisms were high: users expected that the Netflix subject would always capture their tastes and personalities. Algorithmic recommendations were seen as proof, or lack thereof, that this has occurred.

The difficulties in understanding the criteria that inform algorithmic recommendations can also be illustrated with the case of Netflix's so-called alternative genres, that is, categories that are unique to the platform where specific recommendations are included (which have already been discussed in chapter 2). These were the most common categories recommended to interviewees. Alternative genres typically contain traditional generic cues (such as "drama" and "comedy") but problematize them in specific ways. For example, the category "Critically-Acclaimed Binge-Worthy Police TV Dramas" was in 58-year-old university administrator Agustina's profile, and on 25-year-old miscellaneous worker Inés's profile, there was a category for "International Romantic Cheesy Series." For the most part, users found these categories confusing. Carla, herself an audiovisual production college student familiar with genre theory, maintained: "[This category] says it's [about] 'Women Who Rule the Screen,' which is something way too specific. I find that to be funny, but also a bit creepy." Here, the expectation was that genres should be both general (to incorporate variety) and specific (so they can be easily recognized). Many users I interviewed interpreted Netflix's alternative genres as privileging specificity over their expectations of universality.

Interviewees also criticized the lack of clarity in the criteria that tied together the content included in alternative genres and algorithmic bundles. When these criteria were not clear, algorithms lost interpellation power. Elena, a 24-year-old woman who works in a political nongovernmental organization, thus expressed: "The categories [Netflix] creates are plain stupid." Like many other users, Elena felt that these categories lacked common sense as organizing principles. When this occurred, users indicated they lost interest in algorithmic recommendations. Carla concluded: "Thanks for showing me something I know I won't want to watch!"

Finally, a third source of resistance to algorithms centered on users' discomfort or irritation with what they perceived as clear biases in the platform's operation. Users resisted algorithms when they perceived that recommendations were not "personal" but rather guided by an ulterior motive. An expectation of genuineness was thus incorporated into people's relationship with algorithms. This expectation was not met when people noted that popular content was being recommended to them (rather than content "uniquely" selected for each person) and, most often, when the platform recommended its own "Original" productions. Fernanda expressed what really bothered her when this happened: "I feel like they are treating me as part of a mass of users." In a similar manner, Carla said, "I feel it's just too much." Accordingly, Natalia, one of the college students I interviewed, redefined algorithmic hailing as a reminder that Netflix's ultimate desire was to control her through algorithms. Like Fernanda, this made her feel as if she were a "consumer" (in her words) rather than a person.

Some users also resisted the meaning conveyed by certain images, content descriptions, and category names that bundled algorithmic recommendations. These interpellation devices were met with resistance because interviewees felt that aspects of these bundles revealed biases. According to Natalia: "I feel there is a certain bias when it comes to explaining movies, in their synopsis. [Women] are described completely differently than men." The expectation of this user was that Netflix should not use algorithmic bundles to favor sexist structures.

Users resisted not only functional and technical aspects but also the cultural biases inscribed in Netflix's recommendations. These biases were expressed in the constant recommendation of content that users considered stereotypical. Thus, Julieta, a 25-year-old audiovisual producer, complained during the interview: "One went into the 'Latino' catalog and there were only *telenovelas*." Algorithmic biases became evident both in the content that was available and that which was not recommended. Fernando, a 22-year-old lesbian, gay, bisexual, transgender, and intersex (LGBTI) rights activist, explained: "The LGBTI category is nowhere to be found. That category does exist, but Netflix never gives [it to] me, for some strange reason." Fernando's complaint was not that he wished to watch content that was not available on Netflix but rather that the platform was not offering content that should have been suggested to him.

The most common form of infrapolitical resistance was to disregard "failed" recommendations. Elisa, a 40-year-old professional photographer, expressed it in a telling way: "I just ignore [such recommendations]. It's not like I'm going to get into a fight with Netflix!" This assertion reveals the sense of power that is embedded in users' understanding of Netflix's algorithms. Elisa's words suggest that she had learned to work around Netflix rather than aggravating "him." Her statement also suggests that resistance is almost never aimed at transforming the structure of algorithmic interpellation on which Netflix operates but rather at ignoring it.

As García Canclini (2013) noted, ignorance and indifference have been largely disregarded as forms of resistance. But ignoring algorithms is a strategic action for users. Although the analytical focus traditionally has been on much more radical acts, ignorance acknowledges resistance as immanent to power. Ignorance both reproduces and undermines algorithms. In other words, although ignoring algorithms doesn't challenge their authority altogether, it doesn't entirely recognize it either. Insofar as platforms can capture and adjust to it, ignoring algorithms reproduces forms of control. But disregarding algorithmic recommendations also expresses users' ability to "act otherwise" (Giddens 1979, 56), thus manifesting their agency. Ignoring algorithms warrants further consideration in that it embodies the intersection between power and resistance.

Whereas algorithmic recommendations are typically ignored, other actions are taken to let the platform know that the user did not enjoy certain content after it was watched. Carla, the college student cited above, offered an illustration of this practice:

I "thumbed down" *13 Reasons Why*. I thought it was total garbage. I was so angry and I said to myself: "If I do this ['thumb down' the series], [Netflix] will surely take it down." Obviously, that was not the case, but my heart wouldn't let me feel good if I didn't do that.

Carla's comment illustrates the infrapolitical importance of "thumbing down" content or ignoring algorithms. Although she hoped the platform would stop promoting the series, she also knew this would not occur as an immediate result of her act. "Thumbing down" was more about establishing an opinion and affirming her autonomy against the platform's insistence on recommending content that she thought did a disservice to society.

Some users (although a minority) claimed to have employed virtual private network (VPN) applications to reflect Internet protocol (IP) addresses in the United States or Europe and, thus, watch content available in those places. That was the case of Carolina, a college student who used a profile in her uncle's Netflix account:

I always look for something and get the [phrase] "Titles related to . . ." I get the name of the movie but not the actual movie. That infuriates me! Thus, I have VPNs to be able to watch [whichever] Netflix [catalog] from wherever I want, so it becomes more universal, because some movies are only available in other countries.

Carolina's use of the word "universal" reveals her expectation that Netflix should offer the same content everywhere. In return, Netflix has evinced its desire to domesticate users by temporarily blocking the operation of some VPNs.

In this context, many interviewees said they continued to watch content through torrent applications (even those who have Netflix accounts), where several Netflix "Originals" series and films could be found. This also allowed them to have immediate access to content that would take more time to be available on Netflix because of licensing deals.

Despite the criticism, resistance did not translate into attempts to abandon Netflix. Many relativized their criticisms so as not to give up on the platform. Among interviewees, the only reason that led to canceling a subscription was the lack of financial resources after some situations (such as a layoff or change of work). As such, acts of infrapolitical resistance need to be interpreted as an expression of agency. As illustrated throughout this section, users offered accounts of autonomy and empowerment through their resistance to Netflix. At the same time, resistance should be situated within the tensions that underlie the process of mutual domestication. It is difficult to determine how much of users' perception of agency is also designed and engineered by Netflix to domesticate users. For example, the company has consistently emphasized the need to promote a sense of awareness in users to motivate certain consumer behaviors. According to the company's technology blog:

> [An] important element in Netflix' personalization is awareness. We want members to be aware of how we are adapting to their tastes. This not only promotes trust in the system, but encourages members to give feedback that will result in better recommendations. [Amatriain and Basilico 2012a]

For Netflix, a reflexive sense of agency in users does not necessarily predate the domestication of the platform, but it can also be a result of the work of algorithms.

DECONFIGURING THE SPOTIFY USER

Infrapolitical resistance on Spotify centered on how users interpreted the idea that the platform was trying to domesticate them. As Kant (2020) shows, the history of algorithmic recommendation is the history of how technology developers anticipate who will use the platform. In this sense, resistance efforts can be understood as an attempt to "deconfigure" the representation of users that is inscribed in the platform's design and its algorithmic operations. I draw on Woolgar's (1991) notion of "configuration" to theorize this process. According to Woolgar, technology designers develop particular features in artifacts based on the set of actions and behaviors that they assign to putative users. In Woolgar's (1991, 61) words:

> Along with negotiations over who the user might be, comes a set of design (and other) activities which attempt to define and delimit the user's possible actions. Consequently, it is better to say that by setting the parameters for the user's actions, the evolving machine effectively attempts to configure the user.

Users are domesticated in the sense that their identities, roles, and capacities are predefined (or configured) at the design stage. Relating to technology thus implies an invitation to perform a script that contains the actions that are expected of people as ideal users (Akrich 1992; Kant 2020; van Oost 2003). To be sure, user configuration (as theorized by Woolgar) is much more complicated in practice. As Suchman (2012, 56) has noted, "there is no stable designer/user point of view, nor are imaginaries of the user or settings of use inscribed in anything like a complete or coherent form in the object." Moreover, although people are discouraged (or even prohibited) from going "off script," user configuration is never definitive. People can always "interpret, challenge, modify and reject" the actions they are invited to perform (Gillespie 2007, 89).

In the case of Spotify, practices of deconfiguration focused on two specific issues: the kind of individual that the platform invites people to be (a client or paying customer of its services) and the type of relationship with music that it promotes (which privileges music discovery through algorithmic recommendations).

"SECOND-RATE" USERS: AGAINST THE PUSH TO BECOME PAYING CUSTOMERS

A key form of resistance to Spotify was captured by Mayela, a 21-year-old plastic arts student (see figure 6.1). Her rich picture suggested that Spotify's ultimate goal ("*meta*" in Spanish) was not to offer her music but rather to turn her into a paying customer and stop using the app for free. She expressed this by suggesting that Spotify sought to transform her happiness into money (manifested through the dollar symbol next to her fading smile). To achieve its ultimate goal, Spotify exploited her relationship with music, which Mayela represented in various ways: as an addiction ("*adicción musical*"), a form of love (her self-portrait connected the music she was consuming to her heart), and an emotional matter (expressed by the different faces she drew of herself). Mayela then represented Spotify on the upper-right corner as a series of buildings illuminated by an eye-shaped sun (thus pointing to issues of surveillance). Hence the capital S on one of the buildings. Thus, in this drawing, Spotify's exploitation occurs through the transformation of mundane user activities (such as walking around listening to music) into quantitative patterns. The drawing builds on a comparison between natural and urban spaces: quantification stands literally between the beauty of life (walking around surrounded by trees) and a place of homogeneity and dullness (encountering industrial buildings).

During the focus group, Mayela explained her picture as follows: "They [Spotify] gave me the app for free for three months, so I could try it. They wanted me to like it, so I would end up paying and keep it going." In her

Figure 6.1
Mayela's rich picture: Becoming a Spotify paying customer. Drawings by "Mayela" courtesy of the artist.

account, algorithms played a very specific role: "They make you feel heard, valued," she said. But, in her drawing, she seemed ambivalent about the results of this configuration: she stood close to the road that would lead her to Spotify's goal, but she was turning her back on the offer of a "monthly subscription" and focused instead on enjoying her music.

Mayela's picture also captured a more generalized sentiment among several Spotify users, particularly those who employed the free version of the app. These users felt that Spotify offered them a much reduced experience of the platform, compared to that of those who paid a monthly subscription. As Enrique, a 22-year-old business administrator put it: "If you don't pay, you become a second-rate user." Enrique's words suggest that using the app for free implied an essential difference to Spotify in the kind of person he was. This notion of a "second-rate" user is similar to the sentiment that people expressed about being "second-class" Netflix users who were "excluded" by not having access to the same catalog available in the United States.

On Spotify, the sense of exclusion came from not having access to certain features that the app reserved for those who paid for the "Premium" version. For example, users can't listen to music while offline and thus require an Internet connection to access the app. They can't skip more than a few songs each hour and must listen to ads. Certain recommendation features, such as Spotify Radio, are not available. Users experienced these issues as a limitation to the freedom of relating to music the way they wished to, rather than a lack of technological features. As Nina, a 52-year-old audit specialist, described it, Spotify's free version felt "limiting" and "limited" compared to "Premium."

According to Woolgar (1991), user configuration is typically achieved through warnings and blockages that force users to comply with a technological script. In Woolgar's (1991, 79) words, designers make sure that "putative users access the [technology] in the prescribed fashion: by way of preferred (hardware) connections or through a predetermined sequence of [. . .] operations. The user will find other routes barred and warnings posted on the case itself." Spotify users interpreted the free version of the app as a blockage that imposed the obligation to pay a monthly subscription. Consider how Javier, a 20-year-old college student, assessed ads on the platform: "Ads don't bother me as long as they are ads. But [Spotify] is making them increasingly more annoying so you buy the 'Premium' version. Ads are just noise. Thus, I mute them, so I don't have to hear them." For Javier, Spotify's ads were not really ads, that is, instead of promoting something, they operated as blockages meant to annoy people until they ended up paying so they wouldn't have to listen to the ads anymore. Like him, many other users indicated that muting the ads was their preferred infrapolitical response to this attempt at configuration.

In short, although Spotify sought to turn people into "Premium" clients, many claimed the right to be "basic users" (in the words of Pia, a 29-year-old specialist in natural resources management) who had fewer technological features but more freedom in the sense that they could choose how to relate to music.

MUSIC BEYOND ALGORITHMIC DISCOVERY

Another common form of resistance to Spotify among my interviewees focused on the purpose of algorithmic recommendations. Many users interpreted these recommendations as an attempt to define, or configure, their relationship with music by making discovery the primary mode of its consumption. At its core, this interpretation is infrapolitical in that it represents a claim for autonomy. By resisting the principle of the recommendation, users manifested a desire to decide for themselves what music they wanted to listen to. Martín, a college student who had ventured into producing his own music, expressed this in a straightforward manner:

> I hate apps that suggest things to me! Do not tell me what I need to listen to! If I do want to listen to something new, I just ask another real person. And there's something else: I don't like the interface. I feel like I'm in a labyrinth, like I'm getting inside a hole.

Martín thus reacted because of his conviction that what to listen to and when to listen should be his decisions, not Spotify's. The sensation of being "lost" on the app's interface can also be interpreted as his way of saying that everything on the app was configured to decide for him rather than letting him decide for himself.

Users found various ways in which Spotify attempted to configure their relationship with music. They considered that Spotify sought to make music discovery through algorithms into a rule. For example, Javier, the college student cited above, said that Spotify constantly transformed his searches on the platform into song recommendations. Javier explained that many of these searches were relatively insignificant and were not meant to alter the kind of music he expected to hear on the app. However, he felt that Spotify interpreted all his searches as an opportunity to recommend something. In response, Javier indicated he engaged in another infrapolitical practice. Despite the practical inconveniences, he said he preferred to log out of his Spotify account and conduct searches outside the platform so they would not affect algorithms.

This issue expressed a wider problem with the app. For users, discovery was one among many music-consumption practices. "Discovery is a mood in itself," said Sandra, a psychologist, during a focus group. Instead, Spotify worked to turn it into the default. By defining it as a "mood," Sandra also suggested that discovering music needed to result from her own emotions rather than be decided in advance by the platform. Many users thus enacted Spotify as a provider of music (a notion that stressed the importance of people's autonomy in deciding what to listen to) but felt the app's script pushed them to use it as a music-recommendation system instead (thus emphasizing the role of algorithms in making decisions for users).

In addition to the problem of autonomy posed in principle by recommendations, users also criticized them for more practical reasons. As with Netflix's users, people thought that algorithms failed constantly, that is, the algorithms recommended music that was not of interest to them. They also considered that the platform emphasized recommendations of music that typically topped the popularity charts. But many users felt repelled by the concept of popularity in music consumption as a criterion for selection. Instead, they reclaimed their autonomy in deciding what music was relevant to them beyond simply issues of mass approval. Mateo, a 19-year-old college student, captured what he thought was the overall problem of algorithmic recommendations:

> If an algorithm recommends it to me, I don't know if I'm going to listen to it, because it is always giving me music, whether I like it or not. Sometimes I pay attention to it and sometimes I just ignore it. People do have more value [than algorithms,] because they do it [recommend music] personally.

Mateo inverted the assumptions behind conversion practices discussed in chapter 5. Receiving recommendations required the same principle as sharing them with others: knowing the user in a personal manner. He thus felt this could not be achieved by algorithms, whose only job was to recommend music regardless of the situations in the person's life. Mateo also assumed that algorithmic recommendations were motivated by commercial reasons, that is, that the music was suggested to him because someone had paid for this. This violated the assumption of genuineness he expected in music recommendations. Accordingly, he strategically ignored algorithms.

People also saw evidence of Spotify's project of user configuration or domestication in other aspects of the app's operation. For example, many users perceived pressure to consume music through playlists. As noted in chapter 4, playlists are a key in Spotify's current economic project (Bonini and Gandini 2019; Prey, Del Valle, and Zwerwer 2022). It was Martín, the college student, who articulated this idea in the most telling way: "Listening to a playlist is the opposite of listening to an album, because what I like is the narrative pact of listening to an album from beginning to end, and the playlist is like: 'Here's a salad for you'." In this perspective, Spotify not only took away the users' liberty of deciding what music to play by recommending music constantly, but it also betrayed the artistic intention of musicians who had prepared collections of songs to be listened to in a certain order. His metaphor of the salad evoked ideas of mixing music components that were not meant to be removed from their original context (the album). Martín saw evidence of Spotify's attempt to position the playlist as the preferred format for relating to music everywhere in the design of the app.

For all the reasons discussed in this section, many users thought that Spotify's attempt at domesticating them and configuring their relationship with music largely failed. For the most part, they continued to use the app primarily because of its music catalog. Yet, they repeatedly challenged the script that Spotify offered them through infrapolitical actions, such as muting ads, strategically ignoring algorithms, or logging out of their accounts to conduct searches. Rubén, a 39-year-old psychologist, questioned the overall ideal that he thought characterized Spotify's approach to music:

> After five years or more of listening to music on Spotify and YouTube, I said: "How boring!" They let me tailor everything to my liking, but it becomes so individualistic that it gets boring. You can hear everything you want all day. It is the "perfect world."

To illustrate why he thought Spotify lacked most of what made music appealing to him in the first place, Rubén compared it with the traditional use of radio. For Rubén, Spotify misunderstood the significance of radio: rather than wanting to hear songs he had not chosen (as Spotify's "Radio" feature suggested), radio mattered to him because it connected him to strangers who could react to what they were listening to in real time. Accordingly, Rubén said he had begun to listen to some radio stations again. In conclusion, he argued, Spotify's "perfect world" of endless algorithmic recommendation was not so perfect after all.

REVEALING TIKTOK'S POLITICAL PROJECT

Resistance on TikTok could be described as a critical recognition of the app as a political project. In other words, participants in focus groups argued that it was clear to them that TikTok was promoting specific kinds of content, which expressed certain values and worldviews. Drawing on Winner's (1980) classic formulation, this form of resistance highlighted and questioned the politics of the TikTok artifact. It envisioned algorithms as a mechanism to favor certain interests over others. This resistance gave new significance to TikTok: rather than being an entertainment app, it was seen as a technology that worsened existing inequities in the world. In this sense, resistance inverts the rationale that TikTok wishes to convey to users. According to users, TikTok's political project is most obvious in a defining contradiction: it promotes content that should be deleted, while deleting content that should be promoted. In what follows, I examine in more detail this tension and users' responses to it.

PROMOTING WHAT SHOULD BE DELETED

Resistance to TikTok began with the sense that the company favored and enforced a certain worldview. Specifically, participants in focus groups made clear that they did not envision the app as a neutral technology but rather as the expression of a certain way of understanding the world. Valeria, who studied advertising but was unemployed at the time of our conversation, explained this perception with laconic precision: "TikTok is super classist, elitist, and superficial." For Valeria, this meant that TikTok's recommendations often featured white people carrying out activities that showed their wealth or that reinforced stereotypical conceptions of beauty. More broadly, most users agreed that there were clear patterns in the specific body types and social activities that seemed to prevail in TikTok's recommendations. Most arrived at this conclusion through a combination of factors: their own experiences with the app, warnings or complaints posted by other users, and articles in the media that brought the issue of algorithmic bias to their attention.

During focus groups, users assumed an axiological posture to discuss TikTok's worldview. In other words, according to users, there were things that should not be available on the platform. This posture was informed by Costa Ricans' own notions of the public good and the role of algorithmic platforms in shaping it. Users were largely in agreement about the kinds of content they thought that TikTok needed to delete: everything that promoted unrealistic body types and ideals, unhealthy eating habits, racist and xenophobic discourses, stereotypical beauty standards, sexism, the hypersexualization of women, violence, and classist views. More broadly, most agreed that the content displayed on the app lacked ethnic diversity.

According to users, the specific worldview espoused by TikTok was enforced through various means. But more than any other factor, users emphasized the importance of algorithms in this respect. Earlier in this chapter, I discussed how Netflix users recognized an exaggeration bias in the platform's recommendations. In a similar manner, TikTok users argued that the app's agenda became obvious to them when algorithms insistently recommended certain kinds of content. The constant appearance of videos in the recommendations with certain body types and social activities pointed to a pattern of repetition. To capture when this pattern became obvious to them, users employed expressions such as "crossing the line," "a bit excessive," or "not OK anymore." In short, users interpreted the app's insistence on promoting specific content as a bias that revealed the values of the TikTok political project.

More broadly, users felt that the entire TikTok assemblage worked to reinforce such biases. They illustrated this with the filters feature. Daniela, who is 23 years old and recently graduated from college, was among those who most strongly criticized filters:

> The app has filters so that people can modify their face live. I mean, I can record a video and modify my own face, make my nose smaller, make my lips bigger, make my eyes different. In other words, it normalizes the creation of insecurities in people. It is a way of saying: "These are the materials you need [to be beautiful]." This seems very, very dangerous to me.

For Daniela, the problem with filters could be found in how they provided technological means for people to accept certain standards of beauty as a normal part of their lives and how these filters promoted the notion that users needed to do something to achieve those ideals. It was users' perception that the TikTok assemblage then worked to intensify such notions through ritual communication dynamics that enabled a sense of belonging to the app's culture (cf Shifman 2014). Many focus group participants complained that the repetition of playful activities (such as dances, challenges, and chains of content creation) revealed and amplified biases in algorithmic recommendation. Camila, a political scientist who participated in a focus group, used words to describe this situation that are consistent with some work in science and technology studies: "In TikTok there are not only filters but also a script for exposing yourself. TikTok gives you the tools, the hashtags, the songs to talk about the body, to talk about your house, to talk about what you bought." Camila's statement is reminiscent of Akrich's (1992) own view of technology as a script that defines action frameworks for users. For Camila, TikTok's script was made of specific combinations of technological features and cultural ideas through which people learned how to talk about certain issues that, while connecting them to others, reproduced a larger capitalist and neoliberal worldview.

Making this problem worse, users argued, TikTok showed a deliberate effort to not intervene in this state of affairs. Laura, a 19-year-old college student who defined herself as a heavy TikTok user, summarized this view when she claimed: "They [TikTok] just don't do anything about it!" This opinion became most clear when users discussed how they understood content moderation on TikTok. Valentina, a public relations specialist, argued she was convinced that "someone was not doing their job" at moderating content on TikTok. She illustrated this problem with jokes she constantly encountered that, in her opinion, went "over the line." This explanation seemed more probable to users than the possibility that the company's technological procedures to enforce guidelines were failing. If anything, users indicated, TikTok was to blame not for technological failures (thus reconfirming their high opinion of the app's algorithms discussed in chapter 2) but rather for its commitment to virality over moderation. Several users expressed the sense that TikTok used viral trends as an excuse for not deleting content that these users thought should not be on the platform. In the end, this preference reinforced a sort of vicious circle of content that should have been filtered but ended up trending, and then couldn't be deleted because it had gone viral.

To be sure, users did not think that the promotion of sexism, racism, or unhealthy eating habits originated on TikTok or were exclusive problems of the app. They made sure to clarify that they understood these were broader social problems, all of which existed before the creation of TikTok or any other platform. But they also emphasized that these problems manifested in a very particular way on TikTok. The premise behind this belief was a particular notion of TikTok's social power. Valeria, the advertising professional, articulated this idea in a precise manner:

> [On TikTok] there's the "my rich boyfriend check" and there's also its counterpart, something like a "being broke check." But there are people who are literally [poor], who live in overcrowded conditions. *Ay mae!* [Come on!] You [TikTok] cannot normalize this, you cannot normalize [either] the opulence [or] the precarious conditions in which some human beings live. TikTok has many good things that I really appreciate. But I also feel that TikTok's mainstream content is about hegemonic nuances that are excessive.

During our conversation, Valeria specifically contrasted TikTok's algorithmic emphasis on opulence in American society with economic inequities in a country like Costa Rica. In light of this comparison, TikTok's tendency to promote certain kinds of content seemed "excessive." For Valeria, TikTok's power was to normalize hegemonic interpretations of the world. This was done implicitly rather than overtly (hence her use of the word "nuances"). As noted in chapter 2, most users I interviewed thought that TikTok's algorithms were "special" (or, more precisely, particularly "aggressive" in their approach to personalization) compared to other apps. Accordingly, they also argued that the problem of promoting biased content was more acute for TikTok than for the rest of social media platforms. According to users, TikTok could promote biased content more intensively or on a bigger scale precisely because it had fewer mechanisms to control the "aggressiveness" of its algorithms than were available on other platforms.

Despite the threats posed by this situation, users felt they were relatively immune to it. They considered themselves to have the capacity not only to be conscious of algorithmic biases but also to resist their influence. Users extended their concerns to social groups that they deemed more vulnerable. For example, college students who participated in the focus groups seemed particularly worried for teenagers and children who had access to the app. Worries often stemmed from their own experiences with younger siblings. Georgina, a psychology student, wondered whether TikTok had worsened a behavior she had identified in her sister (the one who had motivated her to create a profile on the app):

> I get "TikToks" of girls saying, "Well, this is my food for today," and they eat only two things. And then I got another "TikTok" where the person said: "Stop showing these things because my sister is 13 years old and she sees this content and is going to reproduce it!" I can actually relate to that. My sister is 17 and she has become very aware of how she looks. I don't know if it was the quarantine [lockdown during the COVID-19 pandemic] or TikTok, but she keeps comparing herself to others, noticing how perfect other bodies are.

Parents who participated in the focus groups seemed concerned primarily with their children in their twenties. Just like Georgina stated above, parents argued that they had enough experience in life to resist such forms of influence but wondered whether college life made their children more susceptible to social pressures. A sort of "third-person effect" became evident:

participants expressed concerns mostly in relation to whom they thought were the more vulnerable "others" rather than themselves (Dogruel et al. 2020).

The most common form of resistance used to deal with the content that users thought should not be promoted was to actively employ the "not interested" feature. On one hand, people used this feature strategically to regulate the process of content personalization. In other words, using it was a way to maintain personalization dynamics (as explained in chapter 2). On the other hand, signaling a lack of interest in certain content was an explicit infrapolitical act meant to express something about the person's identity and autonomy. Accordingly, some users valued the "not interested" feature more than the "like" feature. That was the case for Nicolás, a 22-year-old college student, who said that, whereas "likes" were to him a way to signal interest in something, "not interested" was more about letting the platform know his stance on a particular issue. In his opinion, these features had different implications: whereas "likes" had a "trapping effect" (in his words), in the sense that algorithms would continue to offer him the same kind of content over time, "not interested" had a "disengaging" effect, in the sense that it established a distinction between the person and the content.

On other occasions, users went as far as to report content that they thought should be altogether deleted from the platform. Mario, a 23-year-old international relations major, explained the conditions that led him to report content on TikTok: "[I saw] content that made fun of social movements, like Black Lives Matter. That's when I said, '*Mae*, I take offense.' So I reported it right away. It ended up being a political act in itself." Mario acknowledged the political nature of his act. That is, by reporting the content, he expressed his opposition to the kind of society where making fun of certain causes was normal and to the use of TikTok as a platform to enable this society.

DELETING WHAT SHOULD BE PROMOTED

Users denounced a problem that was parallel to the promotion of content they expected to be suppressed on the platform: the app also deleted what should have been promoted. According to users, this tension revealed a double standard in the enforcement of the app's community guidelines. Laura, the college student, said she had encountered content from girls who claimed that their posts had been deleted from the platform because their bodies did not fit certain ideals of beauty. However, Laura noted, "You do find videos of other girls who do fall within stereotypes and who are wearing swimsuits, and no one erased anything." Users like Laura thus felt that TikTok applied community guidelines arbitrarily to favor certain body representations. Instead, these users expected impartiality in the application of rules.

Because focus group conversations took place in the wake of the Black Lives Matter (BLM) movement in the United States, it was by far the example that users discussed the most as the kind of content that should be promoted on the platform. But many users indicated they were convinced that content about BLM was being deleted because of its explicit political nature. Yamila, a biotech engineer, was bothered by TikTok's overall management of BLM. She complained about what she perceived as a bias against Black content creators on the app: "Content from Black people gets buried under what white people post on TikTok," she said. As evidence, users who felt that TikTok either suppressed or downplayed videos about BLM cited their own experiences encountering complaints about this problem, warnings from content creators themselves, and news articles they found in the media. The example of how Costa Rican users interpreted the importance of BLM on TikTok again reveals a deeply held belief in the political power of technology. For users, TikTok should have not deleted posts about this movement, because social media make a difference in changing society. These users commonly assumed that social media platforms should play a role in collective action around issues such as systemic racism, whether in the United States or elsewhere.

Another example of content that users claimed was being deleted deliberately was ethnic diversity. For some users, BLM actually worked as a trigger for paying more attention to issues of diversity on the platform. Valentina, the public relations specialist, said she felt that TikTok's algorithms were "throwing out" (her words for "eliminating") videos with people of color from her "For You" page. She came to this conclusion by specifically noting a lack of Latinx content creators in her recommendations. More broadly, users found counterexamples of content that could replace TikTok's current algorithmic preferences: "healthy lifestyles" rather than "people starving themselves" (as Laura put it), more "diversity" rather than "stereotypes" (in the words of Valentina).

It is important to note that, when it came to illustrate underrepresented issues on TikTok, Costa Rican users relied almost without exception on examples from the United States. Similar to the dynamics discussed in previous chapters, arguing for more diversity on social media or resisting racist algorithmic trends could be interpreted as users' attempts to be a part of larger conversations surrounding technology issues in the world (particularly in the global north). In this way, users could participate from Costa Rica in what they felt was a global conversation and movement, but one that started in one particular country: the United States.

CONCLUDING REMARKS

The study of resistance matters, because it helps mute current "[reassertions of] monolithic accounts of power that tend to downplay or exclude audiences and the significance of the lifeworld" in datafication processes (Livingstone 2019, 171). As an alternative, I have situated the resistance practices of users in Costa Rica in the domain of the infrapolitical. The interest in this form of resistance does not lie necessarily in its ultimate effects of altering domination but rather in the form of critique it performs and how it expresses struggles over meaning and identity. As Scott (1990) has noted, overemphasizing the importance of collective action might precipitate the conclusion that those who do not engage in more explicit resistance practices either lack political perspective or can only express such perspective during moments of disruption. Scott's assertion could also be applied to the case of datafication and algorithms: ordinary platform users don't lack political perspective just because they are not engaged in initiatives for data justice. Infrapolitics is precisely a way to make this perspective more visible.

Through a focus on the infrapolitical, this chapter has invited more nuanced understandings of the relationship between resistance and autonomy than those offered by dominant accounts of datafication (Couldry and Mejias 2019; Danaher 2019). As Calzati (2021, 923) puts it:

> [Data colonialism] seem[s] to conflate two different ideas: one is "predictive analytics" and the other one is "human autonomy." If tech-based monitoring leads to forms of automated guidance, it does not straightforwardly mean that people will lose their autonomy. It is more a negotiation to be at stake, a mutual and case-by-case adaptation.

Taken together, the infrapolitical actions examined in this chapter can be interpreted as claims for identity, autonomy, and dignity that explicitly challenge the passiveness attributed to users in dominant approaches to datafication. But comparing resistance across platforms can also help identify the range of reasons that motivate infrapolitical action. Users resisted a specific blend of commercial and political biases on each platform. In Netflix's case, users vindicated a need to be treated as a person rather than as a consumer "profile" that could easily be targeted through stereotypical and exaggerated interpretations of what it means to watch series and films in Latin America. On Spotify, users asserted their right to define their own relationship with music rather than being "boxed" (as commonly expressed by interviewees) in the kind of framework that benefited the platform both technologically and financially. On TikTok, users' resistance brought to the front (rather than hide) the political project that underlies algorithmic platforms. This became obvious in the kinds of content that users thought should have been deleted and those they felt deserved more presence in algorithmic recommendations.

Across the three cases, users expressed a capacity to identify some of the biases built into algorithms. In Netflix's case, they criticized the platform's tendency to "exaggerate" trends in the production, distribution, and recommendation of content. In particular, they reacted against a perceived tendency to exploit stereotypes about Latin America, genre cues, and even sexist structures. In a similar manner, people criticized Spotify's use of algorithms to promote a particular view of users (as "paying" customers) and one kind of relationship with music consumption. TikTok users also criticized the platform's propensity to promote certain kinds of body types and lifestyles. In short, users perceived a bias against diversity. Identifying these biases was encouraged both by media events and attention to certain issues in public culture and by users' own experiences with these algorithmic platforms. In this sense, people expressed resistance "as a lived experience that is inscribed in [the] meaningful space" of their ordinary practices (Courpasson 2017, 1298).

Infrapolitics is also an ideal framework for making visible the dialectics between compliance and opposition. As Mumby and colleagues (2017, 1161) put it: "resistance and contradiction are frequent bedfellows." In this chapter, I examined specific forms of infrapolitical action that both challenged and reproduced power relations. Ignoring, "thumbing down," "hiding," or reporting algorithmic recommendations illustrate how single and simple acts produce both power and resistance.

Based on the cases analyzed in this chapter, I argue that hegemony remains a much more useful concept than alternatives (including data colonialism) for three reasons. First, it focuses on the dialectical relationship between domination and resistance (Hall 1986; Mumby 1997). Hegemony, as Mumby reminds us, is a continuum, a process of struggle rather than a final state. Second, hegemony points to the significance of culture as a domain of dispute: "hegemony always involves struggle over systems of meaning and the processes by which social reality is framed" (Mumby 1997, 364). And, finally, while authors' use of concepts such as data colonialism typically hides the importance of north-south relations (García Canclini 2020), the notion of hegemony instead points to their significance in how users enact and resist algorithmic platforms. Even if datafication is taking place in both the global north and the south, people experience some of its dynamics in significantly different ways. Netflix users in Costa Rica interpreted catalog differences as a form of exclusion. Many TikTok users also reacted against receiving content that displayed wealth in the US context and the poverty that characterizes parts of countries like Costa Rica.

As discussed in this chapter, the study of resistance touches on the idea of agency. The final chapter addresses the implications of the evidence presented throughout the book for understanding agency in relation to algorithms and datafication.

MUTUAL DOMESTICATION

By considering how users relate to Netflix, Spotify, and TikTok in Costa Rica, the previous chapters have offered evidence to further understand the significance of algorithms in datafication processes. In this final chapter, I take stock of the main lessons derived from this evidence and further consider its implications for thinking about the intersections between algorithms and culture. I begin by revisiting the tension between technologically deterministic accounts of datafication and people's agency, with which this book started. I then further develop the notion of mutual domestication. Going back to a theme introduced in chapter 1, I discuss the book's findings by drawing on the distinction between "algorithms *in* culture" (typically captured with the expression "algorithmic cultures") and "algorithms *as* culture" (how users create and sustain realities by enacting algorithms in particular ways) (Seaver 2017). Building on notions of cyclicity, structuration, and multiplicity, I argue for theorizing these two processes as simultaneous rather than contradictory.

RECOGNIZING THE PURPOSES OF ALGORITHMIC DETERMINISM

Technological determinism—the belief that technology causes social change and is independent of social and political forces (Wyatt 2007)—continues to fascinate commentators and scholars alike. It persists as an underlying assumption in explanations provided when certain artifacts emerge and stabilize. In Wyatt's (2007, 169) words: "technological determinism means that each generation produces a few inventors whose inventions appear to be both the determinants and stepping stones of human development." It could be argued that algorithms (and datafication more broadly) are the inventions of the present generation. They are typically described as autonomous forces that are impacting society in unalterable ways. Accordingly, it has become increasingly common to define the contemporary moment as that of an "algorithmic turn" or an "algorithmic age" (Abiteboul and Dowek 2020).

As Wyatt (2007) shows, the prevalence of various forms of technological determinism is partially a product of how the ideas it embodies speak to our most mundane experiences with technology. Seen in this way, changes in our lives seem to be hardly explained by anything other than ubiquitous technologies. More profoundly, technological determinism also serves a purpose for those who use it (M. R. Smith and Marx 1994). In the case of algorithms, deterministic assumptions have helped frame some of the most profound social, economic, and political transformations of our time. These assumptions have also legitimized a fascination with global technology companies and their operations. Livingstone (2019) argues that the emphasis on the power of algorithms has allowed scholars and commentators to channel valid concerns about the complicated relationships between technology companies and regulatory entities in various countries and regions. In the scholarly domain, it has also served to naturalize the importance of certain methods (such as computational approaches) over others. But even though a great deal of ink and bits have been used to criticize the simplistic narratives of recent popular cultural products that examine the effects of algorithms and datafication (such as Netflix's *The Social Dilemma*), much less has been said about the deterministic premises that still inform much recent scholarly work.

As important and useful as it has been, algorithmic determinism has also come at the price of understanding how people experience processes of datafication in their daily lives. As E. Katz (1980) noted decades ago, when researchers emphasize issues of media and technological power, attention to ordinary people often fades away. Today, there remains a tendency to provide unilateral accounts of power that downplay people's experiences in the analysis of datafication. Boczkowski (2021) already showed how accounts of social media use tend to dehumanize people and deprive them of their most constitutive capacities to act. As Livingstone (2019) aptly notes, this situation has produced a fundamental contradiction: while digital platforms and algorithms become increasingly more central in people's lives around the world, accounts of their experiences with these technologies seem almost dispensable.

In this book, I have argued for responding to the challenges of algorithmic power and the prevalence of deterministic frameworks by critically interrogating people's relationships with digital platforms rather than assuming or erasing these relationships. I have provided empirical evidence to answer the question: what does it mean for people in a Latin American country to live in a datafied society? Whereas most accounts that emphasize algorithmic power tend to strip people away from their agency, I have sought to understand how this agency is enacted through culturally situated relationships between users and algorithms in Costa Rica. Against dominant claims that algorithms have altered, diminished, or even obliterated people's capacity to think, feel, and act, my account has provided a scenario where individuals think, feel, and act with, through, and against algorithms, sometimes enacting their power, sometimes challenging it, often doing both. I have followed a long history of scholarship devoted to examining the everyday lives of people with media technologies, rather than taking them for granted (Martín-Barbero 1993; Rincón and Marroquín 2019). I have argued that the data "behaviors," "traces," and "footprints" of people, often envisioned as fertile terrain for algorithmic manipulation, remain cultural practices and contextual phenomena through and through. Returning to Livingstone's remark about the contradiction in dominant accounts of datafication, I have contended that this kind of empirical investigation is indispensable in light of the prevalence of algorithmic determinism.

FIVE DYNAMICS OF MUTUAL DOMESTICATION

In the previous chapters, I have shown that datafication is a cultural process, endowed with certain dynamics and specificities. In other words, I have argued for considering datafication and people's relationship with algorithms as culture: lived experiences through which people produce, maintain, and transform their realities. This approach allowed me to discuss five dynamics to explain how users in Costa Rica relate to algorithmic platforms. First, I argued for framing *personalization* not exclusively as the willingness to bring personal data to technology companies in order to receive relevant recommendations, but as the establishment of a personal communication relationship with algorithmic platforms. Chapter 2 thus discussed the cases of users configuring their Netflix profiles to reflect aspects of their own personalities, treating Spotify as a person-like being, and undergoing numerous "passages" through which roles and identities for both people and algorithms were assigned and enacted.

Second, I showed that algorithms don't work entirely alone but are rather the product of *integration* dynamics through which they become part of the structures of everyday life. In chapter 3, I examined the variety of sources other than algorithms that users employ to decide what to watch on Netflix. While some of these sources often got entangled with algorithmic recommendations, other users opted to differentiate them as much as possible. I also considered the case of Spotify users who integrated algorithms into their practices based on the specific cultured capacities they wanted to acquire. And, finally, I discussed how people in Costa Rica integrated their experiences with numerous platforms in their relationships with TikTok and its algorithms.

Third, I maintained that user *rituals* allow reproducing the myth of the platformed center, which naturalizes the role of algorithmic platforms as the center of people's lives. In chapter 4, I showed the operation of this myth in various ways: when users turned to Netflix to engage in individual, collective, and hybrid rituals that organized their lives and social relations; when Spotify users engaged in ritual practices to capture moods and emotions through playlists that then worked to assess the affective legitimacy of algorithmic recommendations; and when users ritualized their efforts to deal with boredom through TikTok.

Fourth, I demonstrated how algorithms connect the self and the public through various kinds of *conversion* dynamics. In the case of Netflix, chapter 5 showed how people sought to offer suggestions to others that they felt could not be provided by algorithms, because only humans could go beyond the obvious meaning of cultural products to comment on their personal lives. I also examined how playlists on Spotify functioned as intimate publics through which users could share collective experiences that endowed algorithms with affective meaning. Moreover, I discussed the conversion dynamics on TikTok that users employed to enact the notion of "close friendship" and incorporate algorithms into the process.

Fifth and finally, chapter 6 situated *resistance* efforts in the domain of the infrapolitical, actions that express claims for identity, autonomy, and dignity in relation to datafication and algorithms. I considered users' reactions against certain biases in Netflix's recommendations that materialized a tendency to treat users as stereotypical Latin American consumers. I then showed how Costa Rican users resisted the push to turn algorithms into the center of their experiences with Spotify and the obligation to become paying customers of the app. And I examined how people sought to make TikTok's political project obvious by asking the app to delete what it typically promotes and to promote what it often deletes.

To account for these dynamics, the book employed a methodological approach that looked across the use of three algorithmic platforms. Each chapter thus focused on issues that came to light when the experiences of users of these various platforms were examined empirically. In this way, I sought to counterbalance the emphasis on platforms and "platformization" that tends to characterize the contemporary study of the relationship between algorithms and culture. Furthermore, this comparative method allowed me to identify how these dynamics acquired certain emphases or intensities in specific platforms. In keeping with the notion of mutual domestication, these differences need to be explained as the outcome of both the practices of users and the particular configuration of algorithmic platforms.

For example, rituals proliferated on Netflix. Longstanding practices of using the media as a daily life companion combined with the possibilities to watch content on a variety of devices and in different places. In turn, conversion acquired a certain prominence on Spotify. This was enabled by the affordances of playlisting built into the platform and a deep-seated motivation to share music with others in order to cultivate affect in a public manner. Against the backdrop of a perception that its algorithms were the most "aggressive" of them all, personalization was particularly intense on TikTok. Both integration and resistance characterized the enactment of algorithms on these three platforms, but in somewhat different ways. Integration varied in each platform based not only on algorithmic specificities but also on the situations and demands that users wanted to resolve. Users' resistance emphasized different kinds of algorithmic biases: in some cases, individuals challenged the technological and commercial biases that subtended the operation of algorithms (Spotify); in others, they emphasized political reasons instead (TikTok).

To further understand what it means to live in a datafied society, I have also argued for departing from theorizations that frame agency as an either/ or issue or an all-or-nothing condition (Siles et al. 2022). In other words, I have not sought to replace one form of agency (technological) with another one (human). Instead, the mutual domestication approach espoused in this book situates issues of algorithmic power in the domain of everyday practice. In the remainder of this chapter, I further develop the notion of

mutual domestication and its implications for thinking about the relationship between algorithms and culture.

Although discussed separately, the five dynamics of mutual domestication
that I analyzed in this book need to be understood as being in constant
interaction. Silverstone (1994, 124) emphasized the importance of conceptualizing domestication as "cyclical and dialectical." He elaborated:
"consumption must be seen as a cycle, in which the dependent moments of
consumption [. . .] themselves feed back [. . .] to influence and [. . .] to
define the structure and the pattern of commodification itself" (Silverstone
1994, 124). He evoked the image of a spiral to further explain domestication as a "dialectical movement" (Silverstone 1994, 124). This idea applies
neatly to the case of mutual domestication: its dynamics must be seen as
interdependent, porous, and co-constitutive.

The empirical discussion of the book began with a reconceptualization
of personalization. I used this notion to frame how users sought to establish
communication relationships with algorithmic platforms that interpellate
them. Users responded to algorithmic interpellation by making sense of
platforms as a subject endowed with human-like features. Personalization
then created expectations for other domestication dynamics. It shaped the
performance of rituals and helped establish the idea that platforms operate
as an intermediary of interpersonal relationships that could be developed
through conversion dynamics. Users expected that the "dialogue" they had
established with algorithmic platforms would serve as a basis for making
decisions about which content to choose. In turn, performing certain rituals led to the conviction that it was necessary for users to take care of their
personal profiles in order to communicate with platforms.

Similarly, domestication relied heavily on conversion dynamics. On
some occasions, conversion did not follow other conversion dynamics, as
implied in Silverstone's original analysis, but rather preceded it. Since platforms such as Netflix, Spotify, or TikTok do not advertise in traditional
media in Costa Rica (although they sponsor posts on social media), users
have integrated the work of externalizing public opinions as part of domestication routines. They have also become active in convincing others to
open accounts on these apps (a process that was referred to as "enrollment"

in the case of TikTok). These personal recommendations were a key criterion for many who decided to open a personal profile on these platforms and in establishing rituals for using these platforms. In turn, rituals often culminated in conversion dynamics: users shared suggestions that derived from their rituals in order to define themselves and to further establish the value of interpersonal relationships. Integrating various recommendation sources also triggered conversion dynamics.

Likewise, acts of resistance shaped other dynamics of domestication. As immanent to power, resistance was present in every dimension of the mutual domestication process. Resistance developed as much as successful personalization, integration, rituals, and conversion did. For example, when the expectations derived from algorithmic interpellation were not met, resistance manifested: users assumed that algorithmic platforms should have known them better after they had revealed so much of themselves to interpellating subjects. The frustration with the lack of certain content in Latin America led to new rituals, such as accessing Netflix through virtual private networks (VPNs). This, in turn, allowed users to implement recommendations offered by media outlets or conversations in social media about content not available in Costa Rica and launched new, personalized recommendations. Resistance thus triggered new ways to enact algorithms.

ALGORITHMS *IN* COSTA RICAN CULTURE

As noted in chapter 1, the notion of algorithmic cultures points to how algorithms have come to matter in people's experience of culture. Seen in this way, the previous chapters showed how algorithmic mediations—in Martín-Barbero's (1993) sense of the term—are shaping Costa Rican culture. Algorithmic recommendations condition how users interacted with the "culture machines" of Netflix, Spotify, and TikTok, and how they made sense of their content (Finn 2017).

Users in this study took the infrastructural imperialism of algorithmic platforms as a given of cultural consumption in the present age (Vaidhyanathan 2011). For example, they assumed that individual profiles were the default mode for relating to platforms and engaging with cultural products. Users also naturalized profiling dynamics by framing their consumption practices as part of a dialogue with interpellating subjects in order to personalize their relationship with them. In this way, users interpreted algorithms almost as the utterances of these subjects that revealed how platforms spoke to and hailed them, while also demanding a response from them. Interpellation thus naturalized the use of platforms' features to indicate people's individual preferences and to communicate with interpellating subjects. As the cases of Netflix, Spotify, and TikTok showed, algorithmic interpellation worked successfully when users felt they were offered meaningful recommendations that allowed them to fulfill their expectations and recognize themselves in the content offered to them. This, in turn, reinforced the sense of attachment to algorithmic platforms (often expressed by users themselves through notions such as "addiction").

Another key example of the operation of algorithms *in* culture was their role in naturalizing issues of surveillance. In the case of Costa Rica, users naturalized surveillance in two important ways. First, they linked it to the mythical view of Costa Rica as an intrinsically peaceful and exceptional country. Against the lack of comparable experiences, the surveillance of platforms seemed to be less of a threat to users. Second, they personified algorithmic platforms. This resulted in a common belief that platforms were mediators of personal relationships. In other words, extracting data was almost a favor done to users to better connect them with others.

In essence, algorithmic cultures are about power. Studies that have privileged issues of "platformization" have centered the analysis of power on the intrinsic logics of technology, most notably algorithms. Instead, I framed power through the lens of rituals (Couldry 2003). I argued that algorithms work to naturalize the myth of the platformed center, that is, the notion that algorithmic platforms are the center to which people's lives gravitate. Users sustained this myth through rituals devoted to cultivating practices, moods, and emotions. Algorithmic platforms offered various logics to justify their centrality as the obligatory intermediary in the organization of people's lives. Netflix, for example, legitimized the idea that its algorithms only mirrored the previous actions of its users. Spotify stressed the need to produce, capture, and explore moods through the ritual work of creating playlists. And TikTok offered a temporary solution to the "problem" of boredom through recommendation algorithms that would help users escape the burdens of daily life.

Algorithms were also variously involved in the process of reconnecting the private and the public consumption of platforms in specific ways. A key to understanding the importance of algorithms in this sense was to examine mutually defining notions of self, public, and technology that prevail in Costa Rican society. Users enacted algorithms as an "other" against which they compared their own personal recommendations; an enabler of content discovery; and a crucial and reliable ally in establishing links to others. Despite the differences, algorithms remained crucial in connecting people to others through figures of the public, such as "close friends," "people" in general, or strategic groups of contacts.

As the previous chapters demonstrate, algorithms were key in how users in Costa Rica enacted the local–global distinction. The notion of algorithms *in* culture was expressed in how the people that I studied channeled the cultural aspiration of being part of a global world through the ways they enacted algorithms. To be sure, Costa Rican users consumed series, videos, and music that came from other places, including Latin America (both reggaeton and *Narcos*), South Korea (both K-pop and *Squid Game*), and Spain (both Rosalía and *La Casa de Papel*). But the cultural imaginary that surrounded culture in the United States remained central to their consumption practices. Accordingly, they downplayed mythical views of local culture to favor the notion of inclusion in globalized experiences of culture through algorithms. On these occasions, users reproduced the language and imaginary of "data universalism" (Milan and Treré 2019) and expected that algorithmic platforms would function the same in Costa Rica as they do everywhere else in the world.

Each platform illustrated a different aspect of this process. Netflix users turned to algorithms to further reaffirm their knowledge and appreciation of American cultural products (such as a Hollywood "classic" romantic comedy). They also valued algorithms when these helped them to decide what to watch: they often settled on series and films created in the United States. A common dissatisfaction of both Netflix and Spotify users in the country was not having access to the same catalog these platforms offered to users in the global north (despite paying the same prices for these services). This difference was repeatedly interpreted as an act of exclusion. Furthermore, most of my interviewees created playlists on Spotify with titles in English so that algorithms could help them find an intimate public abroad. Other Costa Rican users noted that TikTok algorithms were able to show them the meaning of queerness (a term they used in English) by reference to meanings that circulated in other parts of the world.

In short, living with algorithms *in* Costa Rica represented an opportunity to participate in a global conversation about series, films, music, and videos in terms that were largely defined in the global north. If the uses of algorithms in Costa Rican look far from "exotic" or different from those of the global north, it is because users work hard to produce these similarities. Inverting T. Porter's (1995) famous argument about quantification, algorithms provided users in this country with a "technology of proximity" that made them feel they belonged in the world. In turn, this dynamic worked to further naturalize the centrality of technology in Costa Rica's economy and national identity (as explained in chapter 1).

ALGORITHMS *AS* COSTA RICAN CULTURE

In his work on television, Silverstone (1994, 112) criticized authors associated with the Frankfurt school for their treatment of culture. In his words:

> [These authors tended] to presume that a cultural logic can be read off from the analysis of industrial logic; to presume a homogeneity of culture which is often more an expression of their own homogenising theories; and they generally fail to acknowledge that culture is plural, that cultures are the products of individual and collective actions, more or less distinctive, more or less authentic, more or less removed from the tentacles of the cultural industry.

By examining algorithms *as* culture, I have sought to counterbalance a similar issue in studies of datafication. Although it captures important aspects of the practices of users, studying datafication by focusing on algorithmic influence *in* culture only (or algorithmic cultures) can also be misleading. Cultures are multidimensional and, as such, the ways in which people relate to media and their contents are conditioned by more than technology. Thus, mutual domestication also requires considering how algorithms are cultures in themselves: they are enacted through practices and rituals; their influence is shaped by users' social, cultural, and professional codes at the national level; and they are resisted and opposed for certain reasons. That is how they acquire meaning and are integrated into daily life. How people build repertoires of criteria for considering what content to choose on platforms such as Netflix and Spotify is also the product of culturally situated processes. The very notion of personalization and the significance of the profile in algorithmic platforms build on the cultural history of individualism.

Academic and journalistic critiques of users in relation to datafication processes and the data extraction practices of companies often miss this point, because they are disconnected from what these technologies actually mean to people and how individuals enact the technologies. Moreover, when algorithmic cultures or data assemblages are empirically examined, authors have often focused on algorithms themselves. As a result, they have emphasized how algorithms can gain fractal capacities to act and shape user practices. The previous chapters have sought to balance out this tendency by noting specifically how users themselves enact data assemblages in ways that can both reproduce and challenge datafication.

Throughout this book, I have used the notion of integration to designate practices through which people enact algorithms as part of cultural reper-toires of resources in ways that allow them to respond to situations in their daily lives. People integrated algorithmic recommendations into matrixes of sources, capacities, and relations based on their sociocultural backgrounds and the material conditions of their digital environments. How Costa Rican users drew on these cultural repertoires was a product of their sociocultural conditions. In short, the structure of everyday life was not swept away with the rise of algorithms in culture.

The enactment of algorithms in Costa Rica was also shaped by the par-ticular ways in which people used culture in this country (Swidler 2001a). As shown in the previous chapters, users constantly turned to mythical views of their country as inherently peaceful, democratic, and egalitarian to relate to algorithms. Accordingly, people "forced" algorithms to comply with local rules of public behavior and not disturb the status quo. They also turned to established views of gender in the country to personify platforms such as Netflix and Spotify as either *él* or *ella*. They appreciated algorithmic recommendations when these allowed them to fulfill the cultural man-date of spending time together with those individuals whom they valued. They strongly protested when algorithms made it seem as if Latin American culture could be reduced exclusively to stereotypes such as drug traffick-ing, soccer, and *telenovelas*. Costa Rican users also interpreted the behavior of TikTok's algorithms as "aggressive" in ways that would not make much sense even in other Latin American countries. In all these ways, algorithms acquired what could be termed as a form of cultural specificity in Costa Rica.

Another example of how users in this country enacted algorithms as cul-ture were resistance practices. The evidence presented throughout the book demonstrates that resistance is highly contextual. In Costa Rica, users often reacted against what they perceived to be forms of exclusion embedded in algorithms. These concerns were meaningful for people located in a Central American country where such issues as lack of diversity, sexism, racism, and eating disorders became common ordinary experiences that demanded a response. North–south relations, often downplayed in dominant analyses of datafication, were crucial in people's relation to algorithmic platforms and the formation of a sense of exclusion typically expressed in such notions as being "second-rate," "second-class," or "basic" users. Put differently, Costa Rican users enacted algorithms by relying on a specific expectation of equality that meant something in particular in Latin American countries in relation to the global north.

The notion of algorithms as culture also makes visible the need to further contextualize key heuristic devices in critical data scholarship. There has been a tendency to treat knowledge of (in the form of imaginaries, folk theories, beliefs, literacies, etc.), affect, and practices with algorithms as universal categories that stand outside of culture and history. The previous chapters have emphasized instead the need to attend to the conditions that make such knowledge, affect, and practices possible in the first place. In other words, there is no single algorithmic imaginary but rather multiple imaginaries that are infused or articulated with cultural values, norms, ideas, rules, and traditions. If, as Bucher (2018, 157) notes, "algorithms help to shape the ways in which we come to know others and ourselves," it is also true that the histories of how we see ourselves and our relationships with others (in places such as Latin America and elsewhere) help shape how we come to know, feel, and act in relation to algorithms. It is through this dual process that algorithms become "eventful" or meaningful (or not) in particular situations (Bucher 2018). This approach provides a much-needed supplement to the growing number of studies that focus on how people think about algorithms by opening up possibilities for understanding why they act the way they do.

THE MULTIPLICITY OF DATAFICATION

The study of algorithms *in* cultures and algorithms *as* cultures has often been framed in oppositional terms. Even Seaver (2017, 10), whose contribution was crucial to elucidate this distinction, argued for the obvious "merits" of the latter approach. He posited the notion of algorithms *as* culture as a solution to understanding the problems entailed by the algorithms *in* culture approach, rather than treating them both as enactments in themselves. In his view, both notions are at odds and are irreconcilable. As an alternative, I argue that when the use of algorithmic platforms is examined empirically, both processes are simultaneous: algorithms are designed to domesticate users and turn them into ideal consumers of algorithmic platforms, but users enact algorithmic recommendations as they incorporate them into their daily lives. The notion of mutual domestication is an attempt to name this process. To further make this case, I rely on Silverstone's work, Giddens' theory of structuration, and a careful consideration of the notion of multiplicity in science and technology studies.

In his work, Silverstone conveyed an idea that is similar to mutual domestication when he theorized domestication as an expression of consumption. He emphasized how individuals were both constrained and free to act in their relationship with television: "We consume and we are consumed. [. . .] In consumption we express *at the same time* and *in the same actions*, both our irredeemable dependence and our creative freedoms as participants in contemporary culture" (Silverstone 1994, 104–105; emphasis added). By theorizing this process as mutual domestication, in this book I have sought to make more conspicuous this constitutive tension in people's relationship with algorithmic platforms, aptly anticipated by Silverstone in the case of television. This requires focusing on both how humans are targeted as *subjects* of domestication (through data extraction processes) how media technologies are *objects* of domestication (through practices of enactment).

Like Silverstone, some authors have found in structuration theory a basis to make similar claims of mutual influence between agents (such as users) and structures (which include media technologies) (Orlikowski 1992; Webster 2011; 2014). Following Giddens, these accounts position users and structures in the duality of their mutual constitution. Technology users rely on structures (including technologies) to enact their agency and, in doing so, they both reproduce and modify such structures.

Orlikowski's (2000) theory of technologies-in-practice blended structuration theory and the notion of enactment to offer an explanation of how people both reproduce and change structures embedded in technology. Orlikowski (1992) argued that structures are not embodied in technology but rather are the product of how users enact them in practice. In other words, people constitute structures through the recurrent use of technologies, through which they "shape the technology structure that shapes their use" (Orlikowski 2000, 407). Technologies are not appropriated but rather enacted through routine, ongoing, and situated practices. Regular and recurrent enactments of technology can lead to institution-alization and reification, although this is not always the case. How users enact technologies-in-practice depends on their own motivations and skills, technologies' affordances, and larger cultural contexts from which people derive imaginaries and understandings. While some enactments reproduce established structures, others can lead to change.

Throughout this book, I have employed the notion of enactment to argue that multiple realities can simultaneously coexist (Law 2004; 2008; Mol 2002). Users can be domesticated into ideal users of algorithmic plat-forms in ways that are consistent with the purposes of data extraction but users also domesticate those algorithms in practice at the same time. Thus in the case of Costa Rican users, algorithms were both a technology of proximity with the global north and a way to reproduce local values, rules, and traditions. This is because enactment efforts are "polyvalent, generat-ing the conditions of possibility for more than one [reality] at the same time" (Omura et al. 2018, 6). This has important implications for how the practices of users are examined: it requires "positioning social actors nei-ther as unwitting dupes who unreflectively reproduce the status quo, nor as individuals who, by virtue of their marginalized status, can create a pristine space of resistance that subverts the dominant order" (Mumby 1997, 366).

Conceptualizing algorithms as multiple thus suggests that datafication is not given but rather is brought into being in various ways through ordi-nary practices. To be sure, I do not mean to suggest that, after consider-ing people's practices, it is possible to conclude that algorithms lack "social power" (Beer 2017). Instead, I have argued that datafication is multiple, an issue that has not been sufficiently discussed in scholarship that accounts primarily for how tech companies and developers of algorithmic systems enact realities. To supplement such dominant approaches, I have provided an empirical account of what people do with algorithms and have consid-ered them as agents that are also involved in the production of algorith-mic power and its resistance. If recent demands for the social sciences and humanities to pay significant attention to algorithms has been met with abundant research (Woolley and Howard 2016), I argue that it is also time for taking seriously how people participate in data assemblages through which agency and power are produced.

Examining the multiplicity and duality of datafication relies on the appropriate recognition of how users enact it by both sustaining and modifying it. Thus, rather than considering either the influence of algorithms in culture or algorithms as cultures only, this approach makes it necessary to study both processes. It acknowledges the possibility of reproducing and resisting datafication. It is precisely in the cyclical interaction and concurrent existence of these two processes that current debates about the intersection of culture and algorithms must be situated. Adapting and extending Silverstone's (1994, 108) dictum: we consume data, we consume through data, we are consumed by data. At the same time. In the same actions.

In this project, I adopted an "ethnographic sensibility" (Star 1999, 383) inspired by the notion of "multi-sited ethnographies" that seek to reveal how meaning is produced and enacted in a multiplicity of settings. I alternated the study of how people related to three algorithmic platforms—Netflix, Spotify, and TikTok—to specifically facilitate incorporating the findings of one case into the study of the others and thus enable a comparative perspective. My research unfolded through five studies conducted over 5 years, which I describe next.

STUDY 1: NETFLIX (2017–2018)

The first study was conducted in 2017 and early 2018 and drew on in-depth interviews with twenty-five Netflix users in Costa Rica. I specifically focused on people who identified themselves as heavy Netflix users. This allowed me to speak with individuals who spent significant time using the platform and whose relationship with it was highly reflexive: that is, they devoted time trying to understand how Netflix operated. I shared a call for participation on social media and selected twenty individuals with different profiles among those who responded. To balance the sample of respondents, I asked interviewees for additional suggestions of people with different backgrounds. The final sample included mostly educated people with a diversity of professional backgrounds who were at different stages in their careers. The age of interviewees ranged between 20 and 53 years old. The final sample also reflected a balance between men (52 percent) and women (48 percent), to reflect patterns identified in existing studies in the country (Red 506 2018). All interviews were conducted in person (between January 2017 and February 2018) and lasted an average of 53 minutes. Most conversations took place in the School of Communication at Universidad de Costa Rica. The interviews were recorded and transcribed in their entirety. I used pseudonyms to protect the identity of interviewees. Interviews were conducted in Spanish. All translations are my own.

To conduct the interviews, I used an adapted version of the "think aloud protocol" (Fonteyn, Kuipers, and Grobe 1993). In this way, I sought to implement the tactic suggested by Seaver (2017, 7) of "treating interviews as fieldwork." I asked participants about their practices and trajectories with Netflix. I also asked them to open their account on a computer, which was projected on a screen so that the research team could simultaneously see the content available and the ways in which people interacted with the platform. I asked informants to describe the particularities of their accounts, their technical configurations, and specific recommendations. I also asked them to reproduce typical appropriation behaviors, discussed specific examples of the content available, and requested explanations of the status of their accounts. Few people interviewed for this study used the term "algorithm" or assigned specific activities to its functioning. And yet forming hypotheses about how recommendations worked was important for them. These users thought they could understand the logic of the recommendation process and thought that their assumptions were correct. These certainties become productive sites to examine how users relate to opaque entities that work to integrate them into the Netflix system and how they gain specific senses of agency through this interaction with algorithms.

Finally, I captured screenshots constantly for the purpose of analyzing them. This made it possible to triangulate data sources, such as verbal descriptions of the interviewees, images, and texts available on the accounts of users. Then I compared accounts from users with descriptions of how Netflix's algorithmic recommendations work that were provided by the company's representatives in the mainstream media and at official outlets. I began analyzing the data by following the main tenets of grounded theory. I carried out open and axial coding individually to identify data patterns, as well as relationships between the patterns. It was through the coding carried out for this first study that I gained an early sense of the five dynamics of mutual domestication that are discussed in this book. I developed some of its first properties (in the case of Netflix) through selective coding.

STUDY 2: SPOTIFY (2018)

The second study was similar to the previous one, but it focused on a different platform. Beuscart, Maillard, and Coavoux (2019) have shown that "heavy" users tend to explore more features in music streaming platforms than casual users do. For this reason, I selected individuals who identified themselves as heavy users of Spotify to identify people with more experience and a deeper understanding of the platform. This strategy allowed me and my research team to interest many individuals who had reflected specifically on how algorithmic recommendations work but could have prevented me from identifying experiences that come from more casual uses of the technology.

I shared a call for participants on social media profiles associated with Universidad de Costa Rica. I selected thirty individuals who responded to this call for interviews. To build this sample, I privileged sociodemographic diversity and thus included fifteen people who identified themselves as men and fifteen who identified themselves as women, aged 19 to 52 years. Participants were mostly educated people with different professional backgrounds. I conducted all the interviews in person between August and November 2018. These interviews lasted for an average of 40 minutes. I recorded these interviews and transcribed them in their entirety. Pseudonyms are used throughout the book to protect the identity of interviewees. (All interviews were conducted in Spanish. Translations are my own.)

I reproduced the experience of study 1 by using an adapted version of the "think aloud protocol." That is, I asked interviewees to open their Spotify accounts on a computer, which was projected on a screen. Those participants who said they accessed Spotify primarily on mobile phones opened their accounts on these devices instead. I then asked interviewees to describe their history of music consumption and typical appropriation practices. We discussed specific instances of content available on their accounts, and I requested explanations of their accounts' configurations. I also inquired into their explanations of how algorithmic recommendations worked. I triangulated data sources by capturing screenshots (of both computer and mobile accounts) and by considering Spotify's discourse about its own services in official outlets. I coded the data inductively in a grounded theory manner (Corbin and Strauss 2015). Three rounds of coding were conducted to develop the main patterns in the data.

STUDY 3: NETFLIX (2019–2020)

The third study drew on twenty-five additional interviews conducted with Netflix users in Costa Rica. This study focused exclusively on Costa Rican Netflix users who identified themselves as women. I began by sharing a call for participants on social media profiles of Universidad de Costa Rica. I sought to foster sociodemographic variety in the construction of the sample. Rather than focusing on heavy users, as the previous Netflix study had done, I specifically sought individuals of different ages, occupations, backgrounds, and experiences with the platform. In this way, I tried to diversify the experiences that served as the basis for my research. I selected a group of twenty-five people for interviews among respondents to this call. The age of interviewees ranged between 19 and 58 years old. Half of the interviewees were younger than 30 years of age and the other half were between 30 and 58 years old.

According to Lobato (2019), access to Netflix is not equally distributed across the world. Although some interviewees preferred not to reveal their approximate monthly income, I would characterize this sample of respondents as middle class. Most of them were educated in a variety of professions.

I conducted all the interviews in person at Universidad de Costa Rica's "Central" campus between March 2019 and February 2020. Conversations lasted for an average of 35 minutes. I recorded these interviews with the approval of each interviewee and transcribed them in their entirety. As with the other studies, I used pseudonyms to protect the identity of respondents, and all interviews were conducted in Spanish. Conversations focused on the history and practices of Netflix use, but they also included discussions of people's backgrounds and social contexts.

I also used a version of the "scrollback technique" advanced by Robards and Lincoln (2017) to supplement the interviews. This method fosters user participation through explanations of certain particularities of their profiles on different platforms. In the words of Robards and Lincoln (2017, 720), this technique "'brings to life' the digital trace, capturing the specific context(s) and contours within which our participants are using [a platform] to make disclosures that we could not intuit without them present." Informants thus become co-analysts of the digital traces they had left over their use trajectory.

With their approval, I asked informants to open their Netflix profiles, which were projected on a screen so that I could see them. I then asked interviewees to describe the main configurations of their profiles on the platform. I also asked them to discuss specific examples of recommendations they had received and requested analytical descriptions of their accounts and practices. I captured videos and screenshots constantly for the purpose of posterior analysis. In this way, I triangulated data sources, such as verbal descriptions of interviewees, videos, images, and texts available on users' profiles.

The rounds of data coding conducted for this particular study helped develop the framework of mutual domestication by specifically identifying empirical properties that had not been noted previously. For example, the notion of algorithmic interpellation became much more obvious in conversations with women who reflected on biased recommendations.

STUDY 4: SPOTIFY (2019–2020)

I opted again for a research design of the qualitative kind to "delve into the workings of assemblages" from the perspective of users (Kitchin and Lauriault 2014, 14). For this study, I supplemented the kind of data I had considered in previous studies to foster more data triangulation. Accordingly, the study employed interviews, focus groups, and rich pictures.

Similar to previous studies, I shared a call for participants on social media profiles associated with Universidad de Costa Rica. I selected thirty individuals among respondents for semi-structured interviews through a criterion strategy that fostered diversity in sociodemographic profiles. This sampling strategy also allowed us to talk with individuals who lived in several provinces of Costa Rica's Central Valley, where most of the population resides. Interviews were conducted at the end of 2019 and beginning of 2020. On average, interviews lasted for approximately 40 minutes.

I also conducted four focus groups with twenty-two additional individuals (aged 18–62 years old) between August and October 2019. I employed the same sampling strategy described in the previous paragraph. That is, I fostered sociodemographic diversity in our sample (although almost all participants have received higher education in various fields). Focus groups were ideal for exploring the social nature of people's understanding of algorithms, that is, how people developed their ideas as they shared them with others (including the researcher). Thus, in addition to gathering data on how individuals accounted for algorithmic recommendations, during the focus groups, I also examined the dialogues, discussions, and collective construction of ideas about algorithms (Cyr 2016). I focused on both the responses to questions about the use of Spotify and the debates that unfolded to answer these questions. Focus groups were also recorded and transcribed.

Finally, I carried out a third research method, namely, rich pictures. Rich pictures are a building block of the so-called soft systems methodology, an approach that emerged in the late 1970s to help actors in conflict reach agreements using a variety of visualization techniques (Checkland 1981). Some of these techniques, primarily rich pictures, can be used in the context of scholarly research as a tool for analyzing "complex situation[s] [and to] provide a space by which participants can negotiate a shared understanding of a context" (Bell, Berg, and Morse 2019, 2). Rich pictures consist of diagrams or drawings made by individuals to graphically represent a specific phenomenon. I employed this technique as a method for making more explicit the unstated and taken-for-granted nature of users' knowledge of algorithms and platforms. I provided participants in focus groups with blank sheets and a set of pens and then asked them to individually draw how they thought that Spotify worked and how it provided them with specific music recommendations. Participants in focus groups explained their own pictures and discussed aspects of other participants' drawings. I then analyzed these pictures by identifying the main patterns in relation to three specific questions: how did users represent Spotify? How did they express a relationship with the platform? How were algorithmic recommendations explained? I used the Bell and Morse (2013) guide to this end and thus coded for patterns in descriptive features and structures (such as use of colors, shapes, thickness, relationships, and arrangements, among others). Coding of the findings from these different methods and sources was carried out to develop the five dynamics of mutual domestication as theoretical constructs.

STUDY 5: TIKTOK (2020)

For the last empirical study, I conducted eight focus groups with thirty-five Costa Rican TikTok users (that is, both people who created videos and those who used the app primarily to watch them) in June and July 2020. I recruited participants through a call for participation that circulated on social media platforms associated with Universidad de Costa Rica. Interested people were asked to fill out an online questionnaire, which allowed me to select potential participants with different sociodemographic characteristics. The final sample included a larger representation of people who identified as women (66 percent) than as men (33 percent). All participants were college students in various local universities or had recently graduated.

They studied majors such as business administration, communication and media studies, engineering, international relations, law, political science, and psychology, among others. Their ages ranged from 18 to 56 years.

Since the research was conducted during the COVID-19 pandemic, all focus groups were conducted on Zoom. Each group consisted of five participants (in addition to the researcher and a research assistant). This proved an ideal number for allowing participants to express their thoughts on a platform such as Zoom. All participants (except one) turned on their cameras during focus groups. Conversations lasted an average of 71 minutes.

After inquiring about participants' personal backgrounds, I asked them to describe their main use practices with TikTok and how these practices had evolved. We then discussed how participants interpreted the operation of algorithms on the app. I examined the answers and debates between participants through which they collectively constructed explanations about the operation of TikTok. To avoid guiding their responses in any particular direction, I did not mention the word "algorithm" in the questions I asked, nor in the explanations of the focus group dynamics. Instead, I asked about people's interpretations of how TikTok recommended content and their thoughts about these recommendations. However, participants explicitly mentioned algorithms in their explanations within 10 minutes into conversations in all the focus groups without exception. When this occurred, I then inquired into their understanding of this term. I recorded the discussions with the approval of participants and transcribed them in their entirety. Again, pseudonyms are used to protect the identity of participants, and I translated from Spanish into English excerpts from these discussions.

The coding of these conversations centered on identifying the main patterns in people's relationships with TikTok's algorithms but also on comparing them with the findings from the previous studies. I used an abductive approach to conduct these two forms of coding. This approach consists of "a continuous process of forming conjectures about a world; conjectures that are shaped by the solutions a researcher already has or can make ready to hand" (Tavory and Timmermans 2014, 28). I considered my previous work on users' relationships with algorithms as part of the "cultivated position" from which I evaluated the findings (Tavory and Timmermans 2014). A final phase of theoretical coding was conducted as an attempt to refine the mutual domestication framework that is developed in this book.

REFERENCES

Abiteboul, Serge, and Gilles Dowek. 2020. *The Age of Algorithms*. Cambridge: Cambridge University Press.

AFP. 2017. "Netflix apuesta por el contenido latinoamericano." Revista Estrategia & Negocios. 2017. https://www.estrategiaynegocios.net/ocio/1095922-330/netflix-apuesta-por-el-contenido-latinoamericano.

Agamben, Giorgio. 2009. *The Signature of All Things: On Method*. New York: Zone Books.

Aguiar, Luis, and Joel Waldfogel. 2018. "Netflix: Global Hegemon or Facilitator of Frictionless Digital Trade?" *Journal of Cultural Economics* 42 (3): 419–445.

Airoldi, Massimo. 2022. *Machine Habitus: Toward a Sociology of Algorithms*. Cambridge: Polity Press.

Akrich, Madeleine. 1992. "The De-Scription of Technical Objects." In *Shaping Technology/Building Society: Studies in Sociotechnical Change*, edited by Wiebe E. Bijker and John Law, 205–224. Inside Technology. Cambridge, MA: MIT Press.

Allen-Robertson, James. 2017. "The Uber Game: Exploring Algorithmic Management and Resistance." *AoIR Selected Papers of Internet Research*, 1–4.

Althusser, Louis. 2014. *On the Reproduction of Capitalism: Ideology and Ideological State Apparatuses*. London: Verso.

Amatriain, Xavier, and Justin Basilico. 2012a. "Netflix Recommendations: Beyond the 5 Stars (Part 1)." Netflix Technology Blog. 2012. https://netflixtechblog.com/netflix-recommendations-beyond-the-5-stars-part-1-55838468f429.

Amatriain, Xavier, and Justin Basilico. 2012b. "Netflix Recommendations: Beyond the 5 Stars (Part 2)." Netflix Technology Blog. 2012. https://netflixtechblog.com/netflix-recommendations-beyond-the-5-stars-part-2-d9b96aa399f5.

Amaya Trujillo, Janny, and Adrien José Charléis Allende. 2018. "Memoria Cultural y Ficción Audiovisual En La Era de La Televisión En Streaming." *Comunicación y Sociedad* 31: 1–27.

Anderson, C. W. 2021. "Fake News Is Not a Virus: On Platforms and Their Effects." *Communication Theory 31* (1): 42–61.

Anderson, Katie Elson. 2020. "Getting Acquainted with Social Networks and Apps: It Is Time to Talk about TikTok." *Library Hi Tech News* 37 (4): 7–12. https://doi.org/10.1108/LHTN-01-2020-0001.

Anderson, Paul Allen. 2015. "Neo-Muzak and the Business of Mood." *Critical Inquiry* 41 (4): 811–840.

Ang, Ien. 1996. *Living Room Wars: Rethinking Media Audiences for a Postmodern World.* London: Routledge.

Argintzona, Jasone. 2020. "Cómo ha evolucionado TikTok en España y Latino-américa." Digimind. 2020. https://blog.digimind.com/es/tendencias/cómo-ha-evoluci onado-tiktok-en-españa-y-américa-latina.

Arnold, Sarah. 2016. "Netflix and the Myth of Choice/Participation/Autonomy." In *The Netflix Effect: Technology and Entertainment in the 21st Century*, edited by Kevin McDonald and Daniel Smith-Rowsey, 49–62. New York: Bloomsbury.

Astudillo, Raquel. 2017. "Netflix aumentará su producción en Latinoamérica 'rápida y agresivamente.'" Datoexpress.cl. 2017. http://www.datoexpress.cl/netflix -aumentara-produccion-latinoamerica-rapida-agresivamente/.

Ávila-Torres, Víctor. 2016. "Making Sense of Acquiring Music in Mexico City." In *Networked Music Cultures*, edited by Raphaël Nowak and Andrew Whelan, 77–93. New York: Palgrave Macmillan.

Bandy, Jack, and Nicholas Diakopoulos. 2020. "#TulsaFlop: A Case Study of Algorithmically-Influenced Collective Action on TikTok." http://arxiv.org/abs /2012.07716.

Baudry, Sandrine. 2012. "Reclaiming Urban Space as Resistance: The Infrapolitics of Gardening." *Revue Française d'Études Américaines* 131 (1): 32–48.

Beer, David. 2017. "The Social Power of Algorithms." *Information, Communication & Society* 20 (1): 1–13.

Beer, David. 2018. *The Data Gaze: Capitalism, Power and Perception.* London: Sage.

Belk, Russell. 2010. "Sharing." *Journal of Consumer Research* 36 (5): 715–734.

Bell, Simon, and Stephen Morse. 2013. "How People Use Rich Pictures to Help Them Think and Act." *Systemic Practice and Action Research* 26 (4): 331–348.

Bell, Simon, Tessa Berg, and Steve Morse. 2019. "Towards an Understanding of Rich Picture Interpretation." *Systemic Practice and Action Research* 32 (60): 1–614.

Benjamin, Ruha. 2019. *Race after Technology: Abolitionist Tools for the New Jim Code.* Cambridge: Polity Press.

Berlant, Lauren. 1997. *The Queen of America Goes to Washington City: Essays on Sex and Citizenship.* Durham, NC: Duke University Press.

Berlant, Lauren. 1998. "Intimacy: A Special Issue." *Critical Inquiry* 24 (2): 281–288.

Berlant, Lauren. 2008. *The Female Complaint: The Unfinished Business of Sentimentality in American Culture*. Durham, NC: Duke University Press.

Berlant, Lauren, and Jay Prosser. 2011. "Life Writing and Intimate Publics: A Conversation with Lauren Berlant." *Biography* 34 (1): 180–187.

Berlant, Lauren, and Michael Warner. 1998. "Sex in Public." *Critical Inquiry* 24 (2): 547–566.

Beuscart, Jean-Samuel, Sisley Maillard, and Samuel Coavoux. 2019. "Les Algorithmes de Recommandation Musicale et l'autonomie de l'auditeur." *Réseaux* 213 (1): 17–47.

Blanco Pérez, Manuel. 2020. "Estética y contexto de los audiovisuales sobre narcotráfico en Latinoamérica en la era Netflix." *Confluenze. Rivista di Studi Iberoamericani* 12 (1): 102–118. https://doi.org/10.6092/issn.2036-0967/11334.

Boczkowski, Pablo J. 1999. "Mutual Shaping of Users and Technologies in a National Virtual Community." *Journal of Communication* 49 (2): 86–108.

Boczkowski, Pablo J. 2021. *Abundance: On the Experience of Living in a World of Information Plenty*. Oxford: Oxford University Press.

Boczkowski, Pablo J., and Eugenia Mitchelstein. 2021. *The Digital Environment*. Cambridge, MA: MIT Press.

Boltanski, Luc, and Laurent Thévenot. 2006. *On Justification: Economies of Worth*. Princeton Studies in Cultural Sociology. Princeton, NJ: Princeton University Press.

Bonini, Tiziano, and Alessandro Gandini. 2019. "'First Week Is Editorial, Second Week Is Algorithmic': Platform Gatekeepers and the Platformization of Music Curation." *Social Media + Society* 5 (4): 1–11.

Brayne, Sarah, and Angèle Christin. 2021. "Technologies of Crime Prediction: The Reception of Algorithms in Policing and Criminal Courts." *Social Problems* 68 (3): 608–624.

Bruns, Axel. 2019. *Are Filter Bubbles Real?* Cambridge: Polity Press.

Bucciferro, Claudia. 2019. "Women and Netflix: Disrupting Traditional Boundaries between Television and Film." *Feminist Media Studies* 19 (7): 1053–1056.

Bucher, Taina. 2018. *If . . . Then: Algorithmic Power and Politics*. Oxford: Oxford University Press.

Burgess, Jean, Peta Mitchell, and Felix Victor Muench. 2019. "Social Media Rituals: The Uses of Celebrity Death in Digital Culture." In *A Networked Self and Birth, Life, Death*, edited by Zizi Papacharissi, 224–239. London: Routledge. https://www.taylorfrancis.com/books/e/9781315202129/chapters/10.4324/9781315202129-14.

Burrell, Jenna. 2016. "How the Machine 'Thinks': Understanding Opacity in Machine Learning Algorithms." *Big Data & Society* 3 (1): 1–12.

Business Wire. 2018. "Spotify Technology S.A. Announces Financial Results for Second Quarter 2018." Business Wire. 2018. https://www.businesswire.com /news/home/20180726005393/en/CORRECTING-and-REPLACING-Spotify -Technology-S.A.-Announces-Financial-Results-for-Second-Quarter-2018.

Butler, Judith. 2016. "Rethinking Vulnerability and Resistance." In *Vulnerability in Resistance*, edited by Judith Butler, Zeynep Gambetti, and Leticia Sabsay, 12–27. Durham, NC: Duke University Press.

Callon, Michel. 1986. "Some Elements of a Sociology of Translation: Domestication of the Scallops and the Fishermen of Saint Brieuc Bay." In *Power, Action and Belief: A New Sociology of Knowledge?*, edited by John Law, 196–233. London: Routledge and Kegan Paul.

Callon, Michel. 2017. *L'emprise Des Marchés: Comprendre Leur Fonctionnement Pour Pouvoir Les Changer*. Paris: La Découverte.

Calzati, Stefano. 2021. "Decolonising 'Data Colonialism'. Propositions for Investigating the Realpolitik of Today's Networked Ecology." *Television & New Media* 22 (8): 914–929.

Cardon, Dominique. 2018. "Le Pouvoir Des Algorithmes." *Pouvoirs* 164 (1): 63–73.

Carey, James W. 1992. *Communication as Culture: Essays on Media and Society*. New York: Routledge.

CEPAL. 2014. "Estado de La Banda Ancha En América Latina y El Caribe 2014." Santiago: Naciones Unidas.

CEPAL. 2016. "Estado de La Banda Ancha En América Latina y El Caribe 2016." Santiago: Naciones Unidas.

CEPAL. 2019. *Panorama Fiscal de América Latina y El Caribe, 2019*. Santiago: CEPAL.

Ceurvels, Matteo. 2020. "TikTok's Corporate Focus on Latin America Is Paying Off." Insider Intelligence. 2020. https://www.emarketer.com/content/tiktoks-corporate -focus-on-latin-america-paying-off.

Chandrashekar, Ashok, Fernando Amat, Justin Basilico, and Tony Jebara. 2017. "Artwork Personalization at Netflix." The Netflix Tech Blog. 2017. https:// netflixtechblog.com/artwork-personalization-c589f074ad76.

Checkland, Peter. 1981. *Systems Thinking, Systems Practice*. Chichester, UK: Wiley.

Cheney-Lippold, John. 2017. *We Are Data: Algorithms and the Making of Our Digital Selves*. New York: New York University Press.

Ciravegna, Luciano. 2012. *Promoting Silicon Valleys in Latin America: Lessons from Costa Rica*. London: Routledge.

Cohn, Jonathan. 2019. *The Burden of Choice: Recommendations, Subversion and Algorithmic Culture*. New Brunswick, NJ: Rutgers University Press.

Contu, Alessia. 2008. "Decaf Resistance: On Misbehavior, Cynicism, and Desire in Liberal Workplaces." *Management Communication Quarterly* 21 (3): 364–379.

Conway, Moody's Analytics, Tractus, and Oxford Economics. 2015. "The World's Most Competitive Cities." Peachtree Corners, GA: Conway.

Corbin, Juliet, and Anselm Strauss. 2015. *Basics of Qualitative Research: Techniques and Procedures for Developing Grounded Theory*. 4th ed. Los Angeles: SAGE.

Cornelio-Marí, Elia Margarita. 2020. "Mexican Melodrama in the Age of Netflix: Algorithms for Cultural Proximity." *Comunicación y Sociedad* 17: 1–27. https://doi.org/10.32870/cys.v2020.7481.

Cotter, Kelley. 2019. "Playing the Visibility Game: How Digital Influencers and Algorithms Negotiate Influence on Instagram." *New Media & Society* 21 (4): 895–913.

Couldry, Nick. 2000. *The Place of Media Power*. London: Routledge.

Couldry, Nick. 2003. *Media Rituals*. London: Routledge.

Couldry, Nick. 2009. "Media Rituals: Beyond Functionalism." In *Media Anthropology*, edited by Eric W. Rothenbuhler and Mihai Coman, 59–69. Thousand Oaks, CA: SAGE.

Couldry, Nick. 2012. *Media, Society, World: Social Theory and Digital Media Practice*. Cambridge: Polity Press.

Couldry, Nick. 2015. "The Myth of 'Us': Digital Networks, Political Change and the Production of Collectivity." *Information, Communication & Society* 18 (6): 608–626.

Couldry, Nick. 2016. "Life with the Media Manifold: Between Freedom and Subjection." In *Politics, Civil Society and Participation: Media and Communications in a Transforming Environment*, edited by Leif Kramp, Nico Carpentier, Andreas Hepp, Richard Kilborn, Risto Kunelius, Hannu Nieminen, Tobias Olsson, Pille Pruulmann-Vengerfeldt, Ilija Tomanić Trivundža, and Simone Tosoni, 25–39. Bremen, Germany: Lumière.

Couldry, Nick, and Ulises A. Mejias. 2019. *The Costs of Connection: How Data Is Colonizing Human Life and Appropriating It for Capitalism*. Stanford, CA: Stanford University Press.

Couldry, Nick, and Ulises Ali Mejias. 2021. "The Decolonial Turn in Data and Technology Research: What Is at Stake and Where Is It Heading?" *Information, Communication & Society*, 1–17.

Courpasson, David. 2017. "Beyond the Hidden/Public Resistance Divide: How Bloggers Defeated a Big Company." *Organization Studies* 38 (9): 1277–1302.

Cozzi, Eugenia. 2019. "El Marginal: Pornografía de La Violencia." Anfibia. 2019. http://revistaanfibia.com/ensayo/pornografia-de-violencia/.

Craig, Robert T. 1999. "Communication Theory as a Field." *Communication Theory* 9 (2): 119–161.

Cuevas Molina, Rafael. 2002. "El Héroe Nacional Costarricense: De Juan Santamaría a Franklin Chang." *Temas de Nuestra América. Revista de Estudios Latinoamericanos* 18 (36): 137–144.

Cuevas Molina, Rafael. 2003. *Tendencias en la dinámica cultural en Costa Rica en el siglo XX.* San José, Costa Rica: EUCR.

Cui, Xi. 2019. "Mediatized Rituals: Understanding the Media in the Age of Deep Mediatization." *International Journal of Communication* 13: 1–14.

Cyr, Jennifer. 2016. "The Pitfalls and Promise of Focus Groups as a Data Collection Method." *Sociological Methods & Research* 45 (2): 231–259.

Cyr, Jennifer. 2019. *Focus Groups for the Social Science Researcher.* Cambridge: Cambridge University Press.

Danaher, John. 2019. "The Ethics of Algorithmic Outsourcing in Everyday Life." In *Algorithmic Regulation*, edited by Karen Yeung and Martin Lodge, 98–118. Oxford: Oxford University Press.

Dayan, Daniel, and Elihu Katz. 1992. *Media Events: The Live Broadcasting of History.* Cambridge, MA: Harvard University Press.

de Melo, Gabriel Borges Vaz, Ana Flávia Machado, and Lucas Resende de Carvalho. 2020. "Music Consumption in Brazil: An Analysis of Streaming Reproductions." *PragMATIZES-Revista Latino-Americana de Estudos Em Cultura* 10 (19): 141–169.

Demont-Heinrich, Christof. 2019. "New Global Music Distribution System, Same Old Linguistic Hegemony?" In *Media Imperialism: Continuity and Change*, edited by Oliver Boyd-Barrett and Tanner Mirrlees, 199–211. Lanham, MD: Rowman & Littlefield.

Dencik, Lina, Arne Hintz, and Jonathan Cable. 2016. "Towards Data Justice? The Ambiguity of Anti-Surveillance Resistance in Political Activism." *Big Data & Society* 3 (2): 1–12. https://doi.org/10.1177/2053951716679678.

Denis, Jérôme, and David Pontille. 2015. "Material Ordering and the Care of Things." *Science, Technology, & Human Values* 40 (3): 338–367.

Denis, Jérôme, and David Pontille. 2020. "Maintenance et Attention à La Fragilité." *Sociologies.* http://journals.openedition.org/sociologies/13936.

DeNora, Tia. 2000. *Music in Everyday Life.* Cambridge: Cambridge University Press.

Deuze, Mark. 2011. "Media Life." *Media, Culture & Society* 33 (1): 137–148.

DeVito, Michael A., Jeremy Birnholtz, Jeffrey T. Hancock, Megan French, and Sunny Liu. 2018. "How People Form Folk Theories of Social Media Feeds and What It Means for How We Study Self-Presentation." In *Proceedings of the 2018 CHI Conference on Human Factors in Computing Systems*, 1–12. Montreal: Association for Computing Machinery.

Dhaenens, Frederik, and Jean Burgess. 2019. "'Press Play for Pride': The Cultural Logics of LGBTQ-Themed Playlists on Spotify." *New Media & Society* 21 (6): 1192–1211.

Dias, Murillo, and Rodrigo Navarro. 2018. "Is Netflix Dominating Brazil?" *International Journal of Business and Management Review* 6 (1): 19–32.

Dias, Ricardo, Daniel Gonçalves, and Manuel J. Fonseca. 2017. "From Manual to Assisted Playlist Creation: A Survey." *Multimedia Tools and Applications* 76 (12): 14375–14403.

Dobson, Amy Shields, Brady Robards, and Nicholas Carah, eds. 2018. *Digital Intimate Publics and Social Media*. Cham, Switzerland: Palgrave Macmillan.

Dogruel, Leyla. 2021. "What Is Algorithm Literacy? A Conceptualization and Challenges Regarding Its Empirical Measurement." In *Algorithms and Communication*, edited by Monika Taddicken and Christina Schumann, 67–93. Digital Communication Research 9. Berlin: Freie Universität Berlin.

Dogruel, Leyla, Dominique Facciorusso, and Birgit Stark. 2020. "'I'm Still the Master of the Machine.' Internet Users' Awareness of Algorithmic Decision-Making and Their Perception of its Effect on their Autonomy." *Information, Communication & Society*, 1–22.

Dourish, Paul. 2016. "Algorithms and Their Others: Algorithmic Culture in Context." *Big Data & Society* 3 (2): 1–11.

Dourish, Paul, and Victoria Bellotti. 1992. "Awareness and Coordination in Shared Workspaces." In *Proceedings of the Conference on Computer-Supported Cooperative Work*, 107–114. Toronto: Association for Computing Machinery.

Dutta, Soumitra, and Irene Mia. 2011. *The Global Information Technology Report 2010–2011*. Vol. 24. Geneva: World Economic Forum.

ECLAC. 2017. *State of Broadband in Latin America and the Caribbean*. Santiago: ECLAC.

Eriksson, Maria, and Anna Johansson. 2017. "Tracking Gendered Streams." *Culture Unbound: Journal of Current Cultural Research* 9 (2): 163–183.

Eriksson, Maria, Rasmus Fleischer, Anna Johansson, Pelle Snickars, and Patrick Vonderau. 2019. *Spotify Teardown: Inside the Black Box of Streaming Music*. Cambridge, MA: MIT Press.

Eslami, Motahhare, Aimee Rickman, Kristen Vaccaro, Amirhossein Aleyasen, Andy Vuong, Karrie Karahalios, Kevin Hamilton, and Christian Sandvig. 2015. "'I Always Assumed That I Wasn't Really That Close to [Her]': Reasoning about Invisible Algorithms in the News Feed." In *Proceedings of ACM Conference on Human Factors in Computing Systems*, 153–162. New York: Association for Computing Machinery.

Espinoza-Rojas, Johan, Ignacio Siles, and Thomas Castelain. 2022. "How Using Various Platforms Shapes Awareness of Algorithms." *Behaviour & Information Technology*: 1–12.

Eubanks, Virginia. 2018. *Automating Inequality: How High-Tech Tools Profile, Police, and Punish the Poor*. New York: St. Martin's Press.

Fagenson, Zachary. 2015. "Spotify Is Eyeing Latin America." Business Insider. 2015. https://www.businessinsider.com/r-spotify-eyes-latin-america-where-growth-on-pace -with-parts-of-europe-2015-5.

Farkas, Johan, Jannick Schou, and Christina Neumayer. 2018. "Platformed Antagonism: Racist Discourses on Fake Muslim Facebook Pages." *Critical Discourse Studies* 15 (5): 463–480.

Ferrari, Fabian, and Mark Graham. 2021. "Fissures in Algorithmic Power: Platforms, Code, and Contestation." *Cultural Studies* 35 (4–5): 814–832.

Ferreira, Gustavo F. C., and R. Wes Harrison. 2012. "From Coffee Beans to Microchips: Export Diversification and Economic Growth in Costa Rica." *Journal of Agricultural and Applied Economics* 44 (4): 517–531.

Finn, Ed. 2017. *What Algorithms Want: Imagination in the Age of Computing*. Cambridge, MA: MIT Press.

Fiore-Gartland, Brittany, and Gina Neff. 2015. "Communication, Mediation, and the Expectations of Data: Data Valences across Health and Wellness Communities." *International Journal of Communication* 9: 1466–1484.

Fisher, Eran. 2022. *Algorithms and Subjectivity: The Subversion of Critical Knowledge*. London: Routledge.

Fiske, John. 1992. "British Cultural Studies and Television." In *Channels of Discourse, Reassembled: Television and Contemporary Criticism*, edited by Robert C. Allen, 214–245. London: Routledge.

Fonteyn, Marsha E., Benjamin Kuipers, and Susan J. Grobe. 1993. "A Description of Think Aloud Method and Protocol Analysis." *Qualitative Health Research* 3 (4): 430–441.

Fortune. 2018. "Latinoamérica impulsa el crecimiento de Spotify." *Fortune* (blog). 2018. https://www.fortuneenespanol.com/finanzas/latinoamerica-crecimiento-spotify/.

Fotopoulou, Aristea. 2019. "Understanding Citizen Data Practices from a Feminist Perspective: Embodiment and the Ethics of Care." In *Citizen Media and Practice: Currents, Connections, Challenges*, edited by Hilde C. Stephansen and Emiliano Treré, 227–242. London: Routledge.

Foucault, Michel. 1997a. "Self Writing." In *Essential Works of Michel Foucault, 1954–1984*, edited by Paul Rabinow, 1. *Ethics: Subjectivity and Truth*, 207–222. New York: New Press.

Foucault, Michel. 1997b. "Technologies of the Self." In *Essential Works of Michel Foucault, 1954–1984*, edited by Paul Rabinow, 1. *Ethics: Subjectivity and Truth*, 223–251. New York: New Press.

Gad, Christopher, Casper Bruun Jensen, and Brit Ross Winthereik. 2015. "Practical Ontology: Worlds in STS and Anthropology." *Nature Culture* 3: 67–86.

Gal, Michal S. 2018. "Algorithmic Challenges to Autonomous Choice." *Michigan Telecommunications Technology Law Review* 25: 59–104.

García Canclini, Néstor. 2013. "¿De Qué Hablamos Cuando Hablamos de Resistencia?" *REVISTARQUIS* 2 (1): 1–23.

García Canclini, Néstor. 2020. *Ciudadanos Reemplazados Por Algoritmos*. Bielefeld, Germany: CALAS.

García Quesada, George. 2020. "Boredom: A Political Issue." In *The Culture of Boredom*, 76–90. Leiden: Brill. https://doi.org/10.1163/9789004427495_005.

Gardiner, Michael E. 2014. "The Multitude Strikes Back? Boredom in an Age of Semiocapitalism." *New Formations* 82: 29–46.

Gentner, Dedre. 2010. "Bootstrapping the Mind: Analogical Processes and Symbol Systems." *Cognitive Science* 34 (5): 752–775.

Gentner, Dedre, and Jose Medina. 1997. "Comparison and the Development of Cognition and Language." *Cognitive Studies* 4 (1): 112–149.

Giddens, Anthony. 1979. *Central Problems in Social Theory: Action, Structure, and Contradiction in Social Analysis*. Vol. 241. Oakland: University of California Press.

Gillespie, Tarleton. 2007. *Wired Shut: Copyright and the Shape of Digital Culture*. Cambridge, MA: MIT Press.

Gillespie, Tarleton. 2014. "The Relevance of Algorithms." In *Media Technologies: Essays on Communication, Materiality and Society*, edited by Tarleton Gillespie, Pablo J. Boczkowski, and Kirsten A. Foot, 167–193. Cambridge, MA: MIT Press.

Gillespie, Tarleton. 2016. "Algorithm." In *Digital Keywords: A Vocabulary of Information Society and Culture*, edited by Benjamin Peters, 18–30. Princeton, NJ: Princeton University Press.

Goldstone, Robert L., Sam Day, and Ji Y. Son. 2010. "Comparison." In *Towards a Theory of Thinking*, edited by Britt M. Glatzeder, Vinod Goel, and Albrecht von Müller, II: 103–122. Heidelberg: Springer.

Gomart, Emilie. 2004. "Surprised by Methadone: In Praise of Drug Substitution Treatment in a French Clinic." *Body & Society* 10 (2–3): 85–110.

Gomart, Emilie, and Antoine Hennion. 1999. "A Sociology of Attachment: Music Amateurs, Drug Users." *Sociological Review* 47 (1): 220–247.

Gómez Ponce, Ariel. 2019. "Hacia Una Concepción Compleja de La Serialización Televisiva En Latinoamérica: Un Análisis Semiótico de Luis Miguel, La Serie." *Dixit* 30 (enero-junio): 22–39. https://doi.org/10.22235/d.v0i30.1744.

Gran, Anne-Britt, Peter Booth, and Taina Bucher. 2021. "To Be or Not to Be Algorithm Aware: A Question of a New Digital Divide?" *Information, Communication & Society* 24 (12): 1779–1796.

Greasley, Alinka E., and Helen M. Prior. 2013. "Mixtapes and Turntablism: DJs' Perspectives on Musical Shape." *Empirical Musicology Review* 8 (1): 23–43.

GSMA. 2016. *Connected Society: Inclusión Digital En América Latina y El Caribe*. London: GSMA.

GSMA. 2020. "La Economía Móvil América Latina 2020." GSMA. https://www
.gsma.com/mobileeconomy/latam-es/.

Guzman, Andrea L., ed. 2018. *Human–Machine Communication: Rethinking Communication, Technology, and Ourselves*. New York: Peter Lang.

Guzman, Andrea L., and Seth C. Lewis. 2020. "Artificial Intelligence and Communication: A Human–Machine Communication Research Agenda." *New Media & Society* 22 (1): 70–86.

Hacking, Ian. 1999. *The Social Construction of What?* Cambridge, MA: Harvard University Press.

Haddon, Leslie. 2017. "The Domestication of Complex Media Repertoires." In *The Media and the Mundane: Communication across Media in Everyday Life*, edited by Kjetil Sandvik, Anne Mette Thorhauge, and Bjarki Valtysson, 17–30. Gothenburg, Sweden: Nordicom.

Hall, Stuart. 1986. "Gramsci's Relevance for the Study of Race and Ethnicity." *Journal of Communication Inquiry* 10 (2): 5–27.

Hallinan, Blake, and Ted Striphas. 2016. "Recommended for You: The Netflix Prize and the Production of Algorithmic Culture." *New Media & Society* 18 (1): 117–137.

Hammer, Rubi, Aharon Bar-Hillel, Tomer Hertz, Daphna Weinshall, and Shaul Hochstein. 2008. "Comparison Processes in Category Learning: From Theory to Behavior." *Brain Research* 1225: 102–118.

Hand, Martin. 2016. "#Boredom: Technology, Acceleration, and Connected Presence in the Social Media Age." In *Boredom Studies Reader*, edited by Michael E. Gardiner and Julian Jason Haladyn, 127–141. London: Routledge.

Hargittai, Eszter, Jonathan Gruber, Teodora Djukaric, Jaelle Fuchs, and Lisa Brombach. 2020. "Black Box Measures? How to Study People's Algorithm Skills." *Information, Communication & Society* 23 (5): 764–775.

Hartmann, Maren. 2013. "From Domestication to Mediated Mobilism." *Mobile Media & Communication* 1 (1): 42–49.

Hartmann, Maren. 2020. "(The Domestication of) Nordic Domestication?" *Nordic Journal of Media Studies* 2 (1): 47–57.

Harvey, Eric. 2014. "Station to Station: The Past, Present, and Future of Streaming Music." Pitchfork. 2014. https://pitchfork.com/features/cover-story/reader/streaming/.

Harvey-Kattou, Liz. 2019. *Contested Identities in Costa Rica: Constructions of the Tico in Literature and Film*. Liverpool: Liverpool University Press.

Hasebrink, Uwe, and Hanna Domeyer. 2012. "Media Repertoires as Patterns of Behaviour and as Meaningful Practices: A Multimethod Approach to Media Use in Converging Media Environments." *Participations* 9 (2): 757–779.

Hearn, Alison, and Sarah Banet-Weiser. 2020. "The Beguiling: Glamour in/as Platformed Cultural Production." *Social Media + Society* 6 (1): 2056305119898779. https://doi.org/10.1177/2056305119898779.

Hecht, John. 2019. "Netflix to Produce 50 Projects in Mexico." Hollywood Reporter. 2019. https://www.hollywoodreporter.com/news/netflix-produce-50-projects-mexico-1185950.

Heeter, Carrie. 1985. "Program Selection with Abundance of Choice: A Process Model." *Human Communication Research* 12 (1): 126–152.

Hennion, Antoine. 2007. "Those Things That Hold Us Together: Taste and Sociology." *Cultural Sociology* 1 (1): 97–114.

Hennion, Antoine. 2013. "D'une Sociologie de La Médiation à Une Pragmatique des Attachements." *Sociologies*. http://journals.openedition.org/sociologies/4353.

Hennion, Antoine. 2016. "From ANT to Pragmatism: A Journey with Bruno Latour at the CSI." *New Literary History* 47 (2–3): 289–308.

Hennion, Antoine. 2017a. "Attachments, You Say? . . . How a Concept Collectively Emerges in One Research Group." *Journal of Cultural Economy* 10 (1): 112–121.

Hennion, Antoine. 2017b. "From Valuation to Instauration: On the Double Pluralism of Values." *Valuation Studies* 5 (1): 69–81.

Hernández Armenta, Mauricio. 2019. "La mujer detrás de Spotify en Latam cuenta como México se volvió especial." Forbes México. 2019. https://www.forbes.com.mx/spotify-en-10-anos-100-millones-de-suscriptores-de-paga/.

Herreros, Sebastián. 2019. *La Regulación Del Comercio Electrónico Transfronterizo En Los Acuerdos Comerciales: Algunas Implicaciones de Política Para América Latina y El Caribe*. Santiago: CEPAL.

Hesmondhalgh, David. 2013. *Why Music Matters*. Oxford: Wiley-Blackwell.

Higgins, Erin Jones. 2017. "The Complexities of Learning Categories through Comparisons." *Psychology of Learning and Motivation* 66: 43–77.

Higueras Ruiz, María José. 2019. "Latin American TV Series Production in Netflix Streaming Platform. Club de Cuervos and La Casa de Las Flores." *Journal of Latin American Communication Research* 7 (1–2): 3–25.

iLifebelt. 2020. "Estudio Tiktok 2020 en Latinoamérica: retos y oportunidades." ILifebelt. 2020. https://ilifebelt.com/estudio-tiktok-2020-en-latinoamerica-retos-y-oportunidades/2020/04/.

INEC. 2020. "Encuesta Nacional de Hogares." San José, Costa Rica: Instituto Nacional de Estadística y Censos.

Infobae. 2019. "Todo lo que tenés que saber sobre TikTok, la app que se convirtió en un fenómeno global." infobae. 2019. https://tendencias/2019/10/21/todo-lo-que-tenes-que-saber-sobre-tiktok-la-app-que-se-convirtio-en-un-fenomeno-global/.

Ingold, Tim. 2018. "One World Anthropology." *HAU: Journal of Ethnographic Theory* 8 (1–2): 158–171.

Introna, Lucas D. 2016. "Algorithms, Governance, and Governmentality: On Governing Academic Writing." *Science, Technology, & Human Values* 41 (1): 17–49.

Iqbal, Mansoor. 2020. "TikTok Revenue and Usage Statistics (2020)." Business of Apps. 2020. https://www.businessofapps.com/data/tik-tok-statistics/.

Jaton, Florian. 2021. *The Constitution of Algorithms*. Cambridge, MA: MIT Press.

John, Nicholas A. 2017. *The Age of Sharing*. Cambridge: Polity.

Johnston, Hank. 2005. "Talking the Walk: Speech Acts and Resistance in Authoritarian Regimes." In *Repression and Mobilization*, edited by Christian Davenport, Hank Johnston, and Carol Mueller, 21:108–137. Minneapolis: University of Minnesota Press.

Jolly, Margaretta. 2011. "Introduction: Life Writing as Intimate Publics." *Biography* 34 (1): v–xi.

Jordán, Valeria, Hernán Galperin, and Wilson Peres Núñez. 2010. "Acelerando La Revolución Digital: Banda Ancha Para América Latina y El Caribe." Santiago: CEPAL.

Kant, Tanya. 2020. *Making It Personal: Algorithmic Personalization, Identity, and Everyday Life*. Oxford: Oxford University Press.

Karaganis, Joe. 2011. *Media Piracy in Emerging Economies*. New York: Social Science Research Council.

Katz, Elihu. 1980. "On Conceptualizing Media Effects." In *Studies in Communication*, edited by Thelma McCormack, 2: 119–141. Greenwich, CT: JAI Press.

Katz, Mark. 2012. *Groove Music: The Art and Culture of the Hip-Hop DJ*. Oxford: Oxford University Press.

Kellogg, Katherine C., Melissa A. Valentine, and Angèle Christin. 2020. "Algorithms at Work: The New Contested Terrain of Control." *Academy of Management Annals* 14 (1): 366–410.

Kennedy, Helen. 2018. "Living with Data: Aligning Data Studies and Data Activism through a Focus on Everyday Experiences of Datafication." *Krisis: Journal of Contemporary Philosophy* 1: 18–30.

Kennedy, Jenny. 2016. "Conceptual Boundaries of Sharing." *Information, Communication & Society* 19 (4): 461–474.

Kitchin, Rob. 2014. *The Data Revolution: Big Data, Open Data, Data Infrastructures and Their Consequences*. London: Sage.

Kitchin, Rob. 2017. "Thinking Critically about and Researching Algorithms." *Information, Communication & Society* 20 (1): 14–29.

Kitchin, Rob, and Tracey P. Lauriault. 2014. "Towards Critical Data Studies: Charting and Unpacking Data Assemblages and Their Work." The Programmable City Working Paper 2. https://papers.ssrn.com/sol3/papers.cfm?abstract_id=2474112.

Klawitter, Erin, and Eszter Hargittai. 2018. "'It's Like Learning a Whole Other Language': The Role of Algorithmic Skills in the Curation of Creative Goods." *International Journal of Communication* 12: 3490–3510.

Kleina, Nilton Cesar Monastier. 2020. "Hora Do TikTok: Análise Exploratória Do Potencial Político Da Rede No Brasil." *Revista UNINTER de Comunicação* 8 (15): 18–34.

Knobloch, Silvia, and Dolf Zillmann. 2002. "Mood Management via the Digital Jukebox." *Journal of Communication* 52 (2): 351–366.

Kokkonen, Tommi. 2017. "Models as Relational Categories." *Science & Education* 26 (7–9): 777–798.

Kotliar, Dan M. 2021. "Who Gets to Choose? On the Socio-Algorithmic Construction of Choice." *Science, Technology, & Human Values 46* (2): 346–375.

Kotras, Baptiste. 2020. "Mass Personalization: Predictive Marketing Algorithms and the Reshaping of Consumer Knowledge." *Big Data & Society* 7 (2): 1–14.

Kushner, Scott. 2013. "The Freelance Translation Machine: Algorithmic Culture and the Invisible Industry." *New Media & Society* 15 (8): 1241–1258. https://doi.org/10.1177/1461444812469597.

Larraín, Felipe, Luis F. López-Calva, and Andrés Rodríguez-Clare. 2001. "Intel: A Case Study of Foreign Direct Investment in Central America." In *Economic Development in Central America*, edited by Felipe Larraín, 1: Growth and Internationalization:165–196. Cambridge, MA: Harvard University Press.

Latinobarómetro. 2018. "Informe 2018." Santiago: Corporación Latinobarómetro.

Law, John. 2004. *After Method: Mess in Social Science Research*. London: Routledge.

Law, John. 2008. "On Sociology and STS." *Sociological Review* 56 (4): 623–649.

Law, John. 2015. "What's Wrong with a One-World World?" *Distinktion: Scandinavian Journal of Social Theory* 16 (1): 126–139.

Lee, Francis, and Lotta Bjorklund Larsen. 2019. "How Should We Theorize Algorithms? Five Ideal Types in Analyzing Algorithmic Normativities." *Big Data & Society* 6 (2): 1–6.

Lee, Francis, Jess Bier, Jeffrey Christensen, Lukas Engelmann, Claes-Fredrik Helgesson, and Robin Williams. 2019. "Algorithms as Folding: Reframing the Analytical Focus." *Big Data & Society* 6 (2): 1–12.

Lena, Jennifer C. 2012. *Banding Together: How Communities Create Genres in Popular Music*. Princeton, NJ: Princeton University Press.

Levy, Karen E. C. 2013. "Relational Big Data." *Stanford Law Review Online* 66: 73–79.

Linke, Gabriele. 2011. "The Public, the Private, and the Intimate: Richard Sennett's and Lauren Berlant's Cultural Criticism in Dialogue." *Biography* 34 (1): 11–24.

Livingstone, Sonia. 2019. "Audiences in an Age of Datafication: Critical Questions for Media Research." *Television & New Media* 20 (2): 170–183.

Livingstone, Sonia, and Peter Lunt. 1994. *Talk on Television: Audience Participation and Public Debate*. Communication and Society. London: Routledge.

Llamas-Rodriguez, Juan. 2020. "Luis Miguel: La Serie, Class-Based Collective Memory, and Streaming Television in Mexico." *JCMS: Journal of Cinema and Media Studies* 59 (3): 137–143.

Lobato, Ramon. 2019. *Netflix Nations: The Geography of Digital Distribution*. New York: New York University Press.

Lomborg, Stine, and Patrick Heiberg Kapsch. 2020. "Decoding Algorithms." *Media, Culture & Society* 42 (5): 745–761.

Lovink, Geert. 2019. *Sad by Design: On Platform Nihilism*. London: Pluto Press.

Lupinacci, Ludmila. 2021. "'Absentmindedly Scrolling through Nothing': Liveness and Compulsory Continuous Connectedness in Social Media." *Media, Culture & Society* 43 (2): 273–290.

Lury, Celia, and Sophie Day. 2019. "Algorithmic Personalization as a Mode of Individuation." *Theory, Culture & Society* 36 (2): 17–37.

Madianou, Mirca, and Daniel Miller. 2013. *Migration and New Media: Transnational Families and Polymedia*. Routledge. https://doi.org/10.4324/9780203154236.

Malaspina, Lucas. 2020. "La Era de TikTok Política, Guerra y Nuevo Lenguaje de Masas | Nueva Sociedad." Nueva Sociedad. 2020. https://nuso.org/articulo/la-era -de-tiktok/.

Marche, Guillaume. 2012a. "Expressivism and Resistance: Graffiti as an Infrapolitical Form of Protest against the War on Terror." *Revue Française d'Études Américaines* 131 (1): 78–96.

Marche, Guillaume. 2012b. "Why Infrapolitics Matters." *Revue Française d'Études Américaines* 1 (131): 3–18.

Martín-Barbero, Jesús. 1986. "Colombia: Prácticas de Comunicación En La Cultura Popular." In *Comunicación Alternativa y Cambio Social*, edited by Máximo Simpson Grinberg, 284–96. Mexico: Premiá.

Martín-Barbero, Jesús. 1993. *Communication, Culture and Hegemony: From the Media to Mediations*. London: SAGE.

Martínez, Laura. 2015. "Spotify abre oficinas en Miami para crecer en América Latina." CNET en Español. 2015. https://www.cnet.com/es/noticias/spotify-abre -oficinas-en-miami-para-crecer-en-america-latina/.

Martínez, Raúl. 2019. "México, el país consentido de Spotify para Latinoamérica." Forbes México. 2019. https://www.forbes.com.mx/mexico-el-pais-consentido -de-spotify-para-latinoamerica/.

Marton, Attila, and Hamid R. Ekbia. 2019. "The Political Gig-Economy: Platformed Work and Labour." In *Fortieth International Conference on Information Systems*, 1–18. Munich. https://aisel.aisnet.org/icis2019/crowds_social/crowds_social/16/.

Mata Hidalgo, Catherine, Luis Oviedo Carballo, and Juan Diego Trejos Solórzano. 2020. *Anatomía de la desigualdad del ingreso en Costa Rica pre Covid-19: investigación de base*. San José, Costa Rica: CONARE-PEN.

Matamoros-Fernández, Ariadna. 2017. "Platformed Racism: The Mediation and Circulation of an Australian Race-Based Controversy on Twitter, Facebook and YouTube." *Information, Communication & Society* 20 (6): 930–946.

Matassi, Mora, Pablo J. Boczkowski, and Eugenia Mitchelstein. 2019. "Domesticating WhatsApp: Family, Friends, Work, and Study in Everyday Communication." *New Media & Society* 21 (10): 2183–2200.

McAlone, Nathan. 2016. "It's Your Fault Netflix Doesn't Have Good Social Features." Business Insider. 2016. https://www.businessinsider.com/netflix-users -dont-want-social-features-2016-2.

McLuhan, Marshall. 1977. "Laws of the Media." *ETC: A Review of General Semantics* 34 (2): 173–179.

Merino, Javier, and Marysabel Huston-Crespo. 2020. "Mia Nygren de Spotify: 'La Ciudad de México es la capital mundial del streaming.'" *CNN* (blog). 2020. https:// cnnespanol.cnn.com/2020/03/05/mia-nygren-de-spotify-la-ciudad-de-mexico-es-la -capital-mundial-del-streaming/.

Miao, Weishan, and Lik Sam Chan. 2021. "Domesticating Gay Apps: An Intersectional Analysis of the Use of Blued among Chinese Gay Men." *Journal of Computer-Mediated Communication* 26 (1): 38–53.

Milan, Stefania. 2019. "Acting on Data(Fication)." In *Citizen Media and Practice: Currents, Connections, Challenges*, edited by Hilde C. Stephansen and Emiliano Treré, 212–225. London: Routledge.

Milan, Stefania, and Emiliano Treré. 2019. "Big Data from the South(s): Beyond Data Universalism." *Television & New Media* 20 (4): 319–335.

Miller, Carolyn R. 1984. "Genre as Social Action." *Quarterly Journal of Speech* 70 (2): 151–167.

Mitchell, Meg Tyler, and Scott Pentzer. 2008. *Costa Rica: A Global Studies Handbook*. Santa Barbara, CA: ABC-CLIO.

Mol, Annemarie. 2002. *The Body Multiple: Ontology in Medical Practice*. Science and Cultural Theory. Durham, NC: Duke University Press.

Monzer, Cristina, Judith Moeller, Natali Helberger, and Sarah Eskens. 2020. "User Perspectives on the News Personalisation Process: Agency, Trust and Utility as Building Blocks." *Digital Journalism* 8 (9): 1142–1162.

Moody, Rebecca. 2018. "Which Countries Pay the Most and Least for Netflix?" Comparitech. 2018. https://www.comparitech.com/blog/vpn-privacy/countries -netflix-cost/.

Moreno, Iago. 2020. "TikTok y los nuevos fascismos." *La Trivial* (blog). 2020. https://latrivial.org/tiktok-y-los-nuevos-fascismos/.

Morley, David. 1986. *Family Television: Cultural Power and Domestic Leisure*. London: Comedia.

Moschetta, Pedro Henrique, and Jorge Vieira. 2018. "Música Na Era Do Streaming: Curadoria e Descoberta Musical No Spotify." *Sociologias* 20 (49): 258–292.

Mosurinjohn, Sharday. 2016. "Overload, Boredom and the Aesthetics of Texting." In *The Boredom Studies Reader: Frameworks and Perspectives*, edited by Michael E. Gardiner and Julian Jason Haladyn, 143–156. London: Routledge.

Mumby, Dennis K. 1997. "The Problem of Hegemony: Rereading Gramsci for Organizational Communication Studies." *Western Journal of Communication* 61 (4): 343–375.

Mumby, Dennis K. 2005. "Theorizing Resistance in Organization Studies: A Dialectical Approach." *Management Communication Quarterly* 19 (1): 19–44.

Mumby, Dennis K., Robyn Thomas, Ignasi Martí, and David Seidl. 2017. "Resistance Redux." *Organization Studies* 38 (9): 1157–1183.

Natale, Simone. 2020. "To Believe in Siri: A Critical Analysis of AI Voice Assistants." *Communicative Figurations Working Papers* 32: 1–17.

Nelimarkka, Matti, Salla-Maaria Laaksonen, Mari Tuokko, and Tarja Valkonen. 2020. "Platformed Interactions: How Social Media Platforms Relate to Candidate-Constituent Interaction during Finnish 2015 Election Campaigning." *Social Media+ Society* 6 (2): 2056305120903856.

Nelson, Roy C. 2009. *Harnessing Globalization: The Promotion of Nontraditional Foreign Direct Investment in Latin America*. University Park, PA: Pennsylvania State University Press.

Netflix. 2017. "Netflix Expande su Inversión en América Latina y Anuncia Nueva Serie Original Filmada Completamente en México, Diablero." Netflix.com. 2017. https://about.netflix.com/es/news/netflix-expands-its-latin-america-investments -announcing-new-original-series-diablero-filmed-entirely-in-mexico.

Nicholson, Brian, and Sundeep Sahay. 2009. "Software Exports Development in Costa Rica: Potential for Policy Reforms." *Information Technology for Development* 15 (1): 4–16.

Nicolau, Anna. 2019. "Spotify quiere replicar su éxito en América Latina en el resto del mundo." Expansión. 2019. https://www.expansion.com/economia-digital /companias/2019/02/01/5c502eed468aeb14338b4643.html.

Noble, Safiya Umoja. 2018. *Algorithms of Oppression: How Search Engines Reinforce Racism*. New York: New York University Press.

Novak, Alison N. 2016. "Narrowcasting, Millennials and the Personalization of Genre in Digital Media." In *The Age of Netflix: Critical Essays on Streaming Media, Digital Delivery and Instant Access*, edited by Cory Barker and Myc Wiatrowski, 162–181. Jefferson, NC: McFarland.

Nowak, Raphaël. 2016. *Consuming Music in the Digital Age: Technologies, Roles and Everyday Life*. New York: Palgrave Macmillan.

Oeldorf-Hirsch, Anne, and German Neubaum. 2021. "What Do We Know about Algorithmic Literacy? The Status Quo and a Research Agenda for a Growing Field." Unpublished manuscript. https://osf.io/preprints/socarxiv/2fd4j.

Omura, Keiichi, Grant Jun Otsuki, Shiho Satsuka, and Atsuro Morita, eds. 2018. *The World Multiple: The Quotidian Politics of Knowing and Generating Entangled Worlds*. London: Routledge.

Ordóñez Ghio, Valeria. 2017. "Netflix alista su mayor apuesta para conquistar México en 2018." Alto Nivel. 2017. https://www.altonivel.com.mx/empresas /apuesta-netflix-conquista-mexico-y-america-latina-2018/.

Orlikowski, Wanda J. 1992. "The Duality of Technology: Rethinking the Concept of Technology in Organizations." *Organization Science* 3 (3): 398–427.

Orlikowski, Wanda J. 2000. "Using Technology and Constituting Structures: A Practice Lens for Studying Technology in Organizations." *Organization Science* 11 (4): 404–428.

Orozco, Guillermo, ed. 2020. *Televisión En Tiempos de Netflix: Una Nueva Oferta Mediática*. Guadalajara: Universidad de Guadalajara.

Pariser, Eli. 2011. *The Filter Bubble: What the Internet Is Hiding from You*. New York: Penguin Press.

Parreño Roldán, Christian. 2013. "Aburrimiento y espacio: Experiencia, modernidad e historia." *REVISTARQUIS* 2: 1–15.

Pasquale, Frank. 2015. *The Black Box Society*. Cambridge, MA: Harvard University Press.

Paus, Eva. 2005. *Foreign Investment, Development, and Globalization: Can Costa Rica Become Ireland?* New York: Palgrave Macmillan.

Penner, Tomaz Affonso, and Joseph D. Straubhaar. 2020. "Netflix Originals and Exclusively Licensed Titles in Brazilian Catalog: A Mapping Producing Countries." *MATRIZes* 14 (1): 125–149.

Pérez-Verdejo, José Manuel, Carlos Adolfo Piña-García, Mario Miguel Ojeda, Alonso Rivera-Lara, and Leonardo Méndez-Morales. 2020. "The Rhythm of Mexico: An Exploratory Data Analysis of Spotify's Top 50." *Journal of Computational Social Science* 4: 147–161.

Perretto, Livia. 2019. "'Brazil Is a Very Important Market for Us,' Reveals TikTok's Community Manager." LABS. 2019. https://labsnews.com/en/articles/technology /brazil-is-a-very-important-market-for-us-reveals-community-manager-tiktok-about -2020-plans/.

Pertierra, Anna Cristina. 2012. "If They Show Prison Break in the United States on a Wednesday, by Thursday It Is Here: Mobile Media Networks in Twenty-First-Century Cuba." *Television & New Media* 13 (5): 399–414.

Pineda, Ricardo, and Carlos Morales. 2015. "Nada puede con Spotify . . . excepto la piratería." Forbes México. 2015. https://www.forbes.com.mx/nada-puede-con -spotify-excepto-la-pirateria/.

Piñón, Juan. 2014. "Reglocalization and the Rise of the Network Cities Media System in Producing Telenovelas for Hemispheric Audiences." *International Journal of Cultural Studies* 17 (6): 655–671.

Poell, Thomas, David B. Nieborg, and Brooke Erin Duffy. 2022. *Platforms and Cultural Production*. Cambridge: Polity Press.

Pomareda, Fabiola. 2021. "La Mitad de Trabajadores del Sector Privado Ganan Menos de ₡400.000 al Mes." Semanario UNIVERSIDAD. September 1, 2021. https://semanariouniversidad.com/pais/la-mitad-de-trabajadores-del-sector-privado -ganan-menos-de-%C2%A2400-000-al-mes/.

Porter, Michael E., and Niels W. Ketelhohn. 2002. "Building a Cluster: Electronics and Information Technology in Costa Rica." *Harvard Business School Case Studies* 703-422: 1–22.

Porter, Theodore M. 1995. *Trust in Numbers: The Pursuit of Objectivity in Science and Public Life*. Princeton, NJ: Princeton University Press.

Powers, Elia. 2017. "My News Feed Is Filtered?" *Digital Journalism* 5 (10): 1315–1335.

Prey, Robert. 2018. "Nothing Personal: Algorithmic Individuation on Music Streaming Platforms." *Media, Culture & Society* 40 (7): 1086–1100.

Prey, Robert, Marc Esteve Del Valle, and Leslie Zwerwer. 2022. "Platform Pop: Disentangling Spotify's Intermediary Role in the Music Industry." *Information, Communication & Society* 25 (1): 74–92.

Prieto, Miriam. 2019. "TikTok, la red china que quiere conquistar el mundo." Expansión. 2019. https://www.expansion.com/economia-digital/companias/2019 /11/08/5dc31167e5fdeaff5b8b457f.html.

Radway, Janice. 1984. *Reading the Romance: Women, Patriarchy, and Popular Literature*. Chapel Hill: University of North Carolina Press.

Rajiva, Mythili, and Stephanie Patrick. 2021. "'This Is What a Feminist Looks Like': Dead Girls and Murderous Boys on Season 1 of Netflix's *You.*" *Television & New Media* 22 (3): 281–298.

Red 506. 2018. "Red 506." San José, Costa Rica: El Financiero.

Reeves, Byron, and Clifford Nass. 1996. *The Media Equation: How People Treat Computers, Television, and New Media Like Real People and Places.* Cambridge: Cambridge University Press.

Resto-Montero, Gabriela. 2016. "The Unstoppable Rise of Reggaeton." Splinter. 2016. https://splinternews.com/the-unstoppable-rise-of-reggaeton-1793854249.

Retana, Camilo. 2011. "Consideraciones acerca del aburrimiento como emoción moral." *Káñina* 35 (2): 179–190.

Ribke, Nahuel. 2021. *Transnational Latin American Television: Genres, Formats and Adaptations.* London: Routledge.

Ricaurte, Paola. 2019. "Data Epistemologies, the Coloniality of Power, and Resistance." *Television & New Media* 20 (4): 350–365.

Rincón, Omar. 2015. "Lo Popular en la Comunicación: Culturas Bastardas+ Culturas Celebrities." In *La Comunicación en Mutación: Remix de Discursos*, edited by Adriana Amado and Omar Rincón, 23–42. Bogotá: Friedrich Ebert Stiftung.

Rincón, Omar, and Amparo Marroquín. 2019. "The Latin American Lo Popular as a Theory of Communication: Ways of Seeing Communication Practices." In *Citizen Media and Practice: Currents, Connections, Challenges*, edited by Hilde C. Stephansen and Emiliano Treré, 42–56. London: Routledge.

Robards, Brady, and Siân Lincoln. 2017. "Uncovering Longitudinal Life Narratives: Scrolling Back on Facebook." *Qualitative Research* 17 (6): 715–730.

Roberge, Jonathan, and Robert Seyfert. 2016. "What Are Algorithmic Cultures?" In *Algorithmic Cultures: Essays on Meaning, Performance and New Technologies*, edited by Robert Seyfert and Jonathan Roberge, 1–25. London: Routledge.

Rodríguez-Arauz, Gloriana, Marisa Mealy, Vanessa Smith, and Joanne DiPlacido. 2013. "Sexual Behavior in Costa Rica and the United States." *International Journal of Intercultural Relations* 13 (1): 48–57.

Rohit, Parimal. 2014. "Personalization, Not House of Cards, Is Netflix Brand." WestSideToday.com. 2014. https://westsidetoday.com/2014/06/17/personalization-house-cards-netflix-brand/.

Rosenblat, Alex. 2018. *Uberland: How Algorithms Are Rewriting the Rules of Work.* Oakland: University of California Press.

Ros Velasco, Josefa. 2017. "Boredom: A Comprehensive Study of the State of Affairs." *Thémata Revista de Filosofía* 56: 171–198. https://doi.org/10.12795/themata.2017.i56.08.

Rothenbuhler, Eric W. 1993. "Argument for a Durkheimian Theory of the Communicative." *Journal of Communication* 43 (3): 158–163.

Rothenbuhler, Eric W. 1998. *Ritual Communication: From Everyday Conversation to Mediated Ceremony*. Thousand Oaks, CA: SAGE.

Rothenbuhler, Eric W. 2006. "Communication as Ritual." In *Communication as . . . : Perspectives on Theory*, edited by Gregory J. Shepherd, Jeffrey St. John, and Ted Striphas, 13–22. Thousand Oaks, CA: SAGE.

Rubio, Francisco. 2012. "El mercado de AL pone a prueba a Netflix." Expansión. 2012. https://expansion.mx/negocios/2012/04/24/el-mercado-de-al-pone-a-prueba-a-netflix.

Sadin, Eric. 2015. *La Vie Algorithmique: Critique de La Raison Numérique*. Paris: Éditions L'échappée.

Salcedo, Moisés. 2019. "TikTok, tu nueva red social favorita." El Universal. 2019. https://www.eluniversal.com.mx/techbit/tiktok-tu-nueva-red-social-favorita.

Salecl, Renata. 2011. *The Tyranny of Choice*. London: Profile Books.

Sandoval, Carlos. 2002. *Threatening Others: Nicaraguans and the Formation of National Identities in Costa Rica*. Athens, OH: Ohio University.

Sandvig, Christian, Kevin Hamilton, Karrie Karahalios, and Cedric Langbort. 2014. "Auditing Algorithms: Research Methods for Detecting Discrimination on Internet Platforms." *Paper presented at the 64th Annual Meeting of the International Communication Association*, 1–23.

Schellewald, Andreas. 2022. "Theorizing 'Stories about Algorithms' as a Mechanism in the Formation and Maintenance of Algorithmic Imaginaries." *Social Media + Society* 8 (1): 1–20.

Schrage, Michael. 2020. *Recommendation Engines*. Cambridge, MA: MIT Press.

Schuilenburg, Marc, and Rik Peeters, eds. 2021. *The Algorithmic Society: Technology, Power, and Knowledge*. London: Routledge.

Schwartz, Sander Andreas, and Martina Skrubbeltrang Mahnke. 2021. "Facebook Use as a Communicative Relation: Exploring the Relation between Facebook Users and the Algorithmic News Feed." *Information, Communication & Society* 24 (7): 1041–1056.

Schwarz, Ori. 2021. *Sociological Theory for Digital Society*. Cambridge: Polity Press.

Scott, James C. 1990. *Domination and the Arts of Resistance: Hidden Transcripts*. New Haven, CT: Yale University Press.

Scott, James C. 2012. "Infrapolitics and Mobilizations: A Response by James C. Scott." *Revue Française d'Études Américaines* 1 (131): 112–117.

Seaver, Nick. 2017. "Algorithms as Culture: Some Tactics for the Ethnography of Algorithmic Systems." *Big Data & Society* 4 (2): 1–12.

Seaver, Nick. 2019a. "Captivating Algorithms: Recommender Systems as Traps." *Journal of Material Culture* 24 (4): 421–436.

Seaver, Nick. 2019b. "Knowing Algorithms." In *DigitalSTS: A Field Guide for Science & Technology Studies*, edited by Janet Vertesi and David Ribes, 412–422. Princeton, NJ: Princeton University Press.

Segura, María Soledad, and Silvio Waisbord. 2019. "Between Data Capitalism and Data Citizenship." *Television & New Media* 20 (4): 412–419.

Shifman, Limor. 2014. *Memes in Digital Culture*. Cambridge, MA: MIT Press.

Siles, Ignacio. 2008. *Por Un Sueño En.Red.Ado. Una Historia de Internet En Costa Rica (1990–2005)*. San José, Costa Rica: Editorial de la Universidad de Costa Rica.

Siles, Ignacio. 2012a. "The Rise of Blogging: Articulation as a Dynamic of Technological Stabilization." *New Media & Society* 14 (5): 781–797.

Siles, Ignacio. 2012b. "Web Technologies of the Self: The Arising of the 'Blogger' Identity." *Journal of Computer-Mediated Communication* 17 (4): 408–421.

Siles, Ignacio. 2017. *Networked Selves: Trajectories of Blogging in the United States and France*. New York: Peter Lang.

Siles, Ignacio. 2020. *A Transnational History of the Internet in Central America, 1985–2000: Networks, Integration, and Development*. New York: Palgrave Macmillan.

Siles, Ignacio, and Pablo J. Boczkowski. 2012. "At the Intersection of Content and Materiality: A Texto-Material Perspective on Agency in the Use of Media Technologies." *Communication Theory* 22 (3): 227–249.

Siles, Ignacio, Johan Espinoza-Rojas, and Andrés Méndez. 2016. "¿El Silicon Valley Latinoamericano?: La Producción de Tecnología de Comunicación En Costa Rica." *Anuario de Estudios Centroamericanos* 42: 411–431.

Siles, Ignacio, Johan Espinoza-Rojas, Adrián Naranjo, and María Fernanda Tristán. 2019a. "The Mutual Domestication of Users and Algorithmic Recommendations on Netflix." *Communication, Culture & Critique* 12 (4): 499–518.

Siles, Ignacio, Edgar Gómez-Cruz, and Paola Ricaurte. 2022. "Toward a Popular Theory of Algorithms." *Popular Communication*: 1–14.

Siles, Ignacio, Andrés Segura-Castillo, Mónica Sancho, and Ricardo Solís-Quesada. 2019b. "Genres as Social Affect: Cultivating Moods and Emotions through Playlists on Spotify." *Social Media + Society* 5 (2): 1–11.

Siles, Ignacio, Andrés Segura-Castillo, Ricardo Solís-Quesada, and Mónica Sancho. 2020. "Folk Theories of Algorithmic Recommendations on Spotify: Enacting Data Assemblages in the Global South." *Big Data & Society* 7 (1): 1–15.

Silverstone, Roger. 1994. *Television and Everyday Life*. London: Routledge.

Silverstone, Roger. 2006. "Domesticating Domestication. Reflections on the Life of a Concept." In *Domestication of Media and Technology*, edited by Thomas Berker, Maren Hartmann, Yves Punie, and Katie Ward, 229–248. Maidenhead, UK: Open University Press.

Siri, Laura. 2016. "El Rol de Netflix En El Ecosistema de Medios y Telecomunicaciones. ¿El Fin de La Televisión y Del Cine?" *Hipertextos* 4 (5): 47–109.

Smith, Aaron, Kyley McGeeney, Maeve Duggan, Lee Rainie, and Scott Keeter. 2015. "U.S. Smartphone Use in 2015." Washington, DC: Pew Research Center. https://www.pewresearch.org/internet/2015/04/01/us-smartphone-use-in-2015/.

Smith, Merritt Roe, and Leo Marx. 1994. *Does Technology Drive History? The Dilemma of Technological Determinism*. Cambridge, MA: MIT Press.

Smith, Paul Julian. 2019. *Multiplatform Media in Mexico: Growth and Change since 2010*. New York: Palgrave Macmillan.

Solsman, Joan E. 2019. "Netflix Finally Spilled How Many Members It Has Region by Region." CNET. 2019. https://www.cnet.com/news/netflix-finally-spilled-how-many-members-it-has-region-by-region/.

Soto Morales, Eugenia. 2014. "Costa Rica destaca como primer exportador de alta tecnología en Latinoamérica y el cuarto en el mundo." *El Financiero* 2014. https://www.elfinancierocr.com/economia-y-politica/costa-rica-destaca-como-primer-exportador-de-alta-tecnologia-en-latinoamerica-y-el-cuarto-en-el-mundo/II67XMKQR5CCVB3GFSY6CGMIMI/story/.

Spar, Debora. 1998. "Attracting High Technology Investment: Intel's Costa Rican Plant." FIAS Occasional Paper 11. Washington, DC: World Bank.

Spotify. 2018. "The Top Songs, Artists, Playlists, and Podcasts of 2018." Spotify. https://newsroom.spotify.com/2018-12-04/the-top-songs-artists-playlists-and-podcasts-of-2018/.

Spotify. 2019a. "6 Questions (and Answers) with Mia Nygren, Managing Director, Spotify Latin America." Spotify. https://newsroom.spotify.com/2019-07-23/6-questions-and-answers-with-mia-nygren-managing-director-spotify-latin-america/.

Spotify. 2019b. "Create Playlists." Spotify. https://support.spotify.com/us/article/create-playlists/.

Srnicek, Nick. 2016. *Platform Capitalism*. Cambridge: Polity Press.

Star, Susan Leigh. 1999. "The Ethnography of Infrastructure." *American Behavioral Scientist* 43 (3): 377–391.

Statista. 2021a. "Netflix's Annual Revenue in Latin America from 2017 to 2020." Statista.com. 2021. https://www.statista.com/statistics/1088546/annual-revenue-netflix-latin-america/.

Statista. 2021b. "Spotify: Monthly Active Users in Latin America 2021." Statista .com. 2021. https://www.statista.com/statistics/813870/spotify-monthly-active-users -quarter/.

Straubhaar, Joseph D., Deborah Castro, Luiz Guilherme Duarte, and Jeremiah Spence. 2019. "Class, Pay TV Access and Netflix in Latin America: Transformation within a Digital Divide." *Critical Studies in Television* 14 (2): 233–254. https:// doi.org/10.1177/1749602019837793.

Streeter, Thomas. 2010. *The Net Effect: Romanticism, Capitalism, and the Internet.* New York: New York University Press.

Strengers, Yolande, and Jenny Kennedy. 2020. *The Smart Wife: Why Siri, Alexa, and Other Smart Home Devices Need a Feminist Reboot.* Cambridge, MA: MIT Press.

Striphas, Ted. 2015. "Algorithmic Culture." *European Journal of Cultural Studies* 18 (4–5): 395–412.

Suchman, Lucy. 2012. "Configuration." In *Inventive Methods: The Happening of the Social*, edited by Celia Lury and Nina Wakeford, 48–60. London: Routledge.

Sued, Gabriela Elisa, María Concepción Castillo-González, Claudia Pedraza, Doris-milda Flores-Márquez, Sofía Álamo, María Ortiz, Nohemí Lugo, and Rosa Elba Arroyo. 2022. "Vernacular Visibility and Algorithmic Resistance in the Public Expression of Latin American Feminism." *Media International Australia 183* (1): 60–76.

Svendsen, Lars. 2005. *A Philosophy of Boredom.* London: Reaktion Books.

Swart, Joëlle. 2021. "Experiencing Algorithms: How Young People Understand, Feel about, and Engage with Algorithmic News Selection on Social Media" 7 (2): 1–11.

Swidler, Ann. 1986. "Culture in Action: Symbols and Strategies." *American Sociological Review* 51 (2): 273–286.

Swidler, Ann. 2001a. *Talk of Love: How Culture Matters.* Chicago: University of Chicago Press.

Swidler, Ann. 2001b. "What Anchors Cultural Practices." In *The Practice Turn in Contemporary Theory*, edited by Theodore R. Schatzki, Karin Knorr Cetina, and Eike von Savigny, 83–101. London: Routledge.

Taneja, Harsh, James G. Webster, Edward C. Malthouse, and Thomas B. Ksiazek. 2012. "Media Consumption across Platforms: Identifying User-Defined Repertoires." *New Media & Society* 14 (6): 951–68. https://doi.org/10.1177/1461444811436146.

Tavory, Iddo, and Stefan Timmermans. 2014. *Abductive Analysis: Theorizing Qualitative Research.* Chicago: University of Chicago Press.

Taylor, Charles. 1989. *Sources of the Self: The Making of the Modern Identity.* Cambridge, MA: Harvard University Press.

Toohey, Peter. 2019. "Is It a Good Thing to Be Bored?" In *Boredom Is in Your Mind*, edited by Josefa Ros Velasco, 1–10. Cham, Switzerland: Springer.

Turkle, Sherry. 2015. *Reclaiming Conversation: The Power of Talk in a Digital Age.* New York: Penguin Press.

Vaidhyanathan, Siva. 2011. *The Googlization of Everything (And Why We Should Worry).* Berkeley: University of California Press.

Vallee, Mickey. 2020. "Doing Nothing Does Something: Embodiment and Data in the COVID-19 Pandemic." *Big Data & Society* 7 (1): 1–12. https://doi.org/10.1177/2053951720933930.

van Dijck, José, Thomas Poell, and Ester de Waal. 2018. *The Platform Society: Public Values in a Connective World.* Oxford: Oxford University Press.

van Oost, Ellen. 2003. "Materialized Gender: How Shavers Configure the User's Femininity and Masculinity." In *How Users Matter: The Co-Construction of Users and Technology,* edited by Nelly Oudshoorn and Trevor Pinch, 193–208. Cambridge, MA: MIT Press.

Varise, Franco. 2011. "Más Allá de La Televisión: Cuevana Sacude La Forma de Ver." *La Nación,* 2011. https://www.lanacion.com.ar/espectaculos/cuevana-sacude-la-forma-de-ver-nid1375216/.

Velkova, Julia, and Anne Kaun. 2021. "Algorithmic Resistance: Media Practices and the Politics of Repair." *Information, Communication & Society* 24 (4): 523–540.

Verbeek, Peter-Paul. 2005. *What Things Do: Philosophical Reflections on Technology, Agency, and Design.* University Park, PA: Pennsylvania State University Press.

Vogel, Thomas T. 1998. "Costa Rica Succeeds in Attracting Investment from High-Tech Firms." *Wall Street Journal,* 1998, sec. Tech Center. https://www.wsj.com/articles/SB8914694306530500.

Wachter-Boettcher, Sara. 2017. *Technically Wrong: Sexist Apps, Biased Algorithms, and Other Threats of Toxic Tech.* New York: W. W. Norton & Company.

Waisbord, Silvio. 2019. *Communication: A Post-Discipline.* Cambridge: Polity Press.

Wajcman, Judy. 2004. *Technofeminism.* Cambridge: Polity.

Webster, James G. 2011. "The Duality of Media: A Structurational Theory of Public Attention." *Communication Theory* 21 (1): 43–66.

Webster, James G. 2014. *The Marketplace of Attention: How Audiences Take Shape in a Digital Age.* Cambridge, MA: MIT Press.

Webster, James G., and Jacob J. Wakshlag. 1983. "A Theory of Television Program Choice." *Communication Research* 10 (4): 430–446. https://doi.org/10.1177/009365083010004002.

Werner, Ann. 2020. "Organizing Music, Organizing Gender: Algorithmic Culture and Spotify Recommendations." *Popular Communication* 18 (1): 78–90.

Williams, Raymond. 2001. *The Long Revolution.* Peterborough, Ontario: Broadview Press.

Winner, Langdon. 1980. "Do Artifacts Have Politics?" *Daedalus* 109 (1): 121–136.

Woolgar, Steve. 1991. "Configuring the User: The Case of Usability Trials." In *A Sociology of Monsters: Essays on Power, Technology and Domination*, edited by John Law, 57–99. London: Routledge.

Woolley, Samuel C., and Philip N. Howard. 2016. "Political Communication, Computational Propaganda, and Autonomous Agents: Introduction." *International Journal of Communication* 10: 4882–4890.

Woyciekowski, Rafael, and Eduardo Zilles Borba. 2020. "Mediações Algorítmicas No Spotify: A Experiência Personalizada Do Usuário Na Playlist Descobertas Da Semana." *ICom* 3 (1): 88–115.

Wyatt, Sally. 2007. "Technological Determinism Is Dead; Long Live Technological Determinism." In *The Handbook of Science and Technology Studies,* Third Edition, edited by Edward J. Hackett, Olga Amsterdamska, Michael Lynch, and Judy Wajcman, 165–180. Cambridge, MA: MIT Press.

Yeung, Karen. 2018. "Algorithmic Regulation: A Critical Interrogation." *Regulation & Governance* 12 (4): 505–523.

Ytre-Arne, Brita, and Hallvard Moe. 2021a. "Doomscrolling, Monitoring and Avoiding: News Use in COVID-19 Pandemic Lockdown." *Journalism Studies* 22 (13): 1739–1755.

Ytre-Arne, Brita, and Hallvard Moe. 2021b. "Folk Theories of Algorithms: Understanding Digital Irritation." *Media, Culture & Society* 43 (5): 807–824. https://doi.org/10.1177/0163443720972314.

Yúdice, George. 2012. "New Social and Business Models in Latin American Musics." In *Consumer Culture in Latin America*, edited by John Sinclair and Anna Cristina Pertierra, 17–33. New York: Palgrave Macmillan.

Ziewitz, Malte. 2017. "A Not Quite Random Walk: Experimenting with the Ethnomethods of the Algorithm." *Big Data & Society* 4 (2): 1–13.

Zillmann, Dolf. 2000. "Mood Management in the Context of Selective Exposure Theory." *Annals of the International Communication Association* 23 (1): 103–123.

Zuboff, Shoshana. 2019. *The Age of Surveillance Capitalism: The Fight for a Human Future at the New Frontier of Power.* New York: PublicAffairs.

INDEX